Blockchain in Supply Chain Digital Transformation

Editor

Dr. Trevor Clohessy
Atlantic Technological University
Renmore, Galway, Ireland

CRC Press is an imprint of the
Taylor & Francis Group, an **informa** business
A SCIENCE PUBLISHERS BOOK

Cover credit: "Vectorjuice - Freepik.com". This cover has been designed using image from Freepik.com

First edition published 2023
by CRC Press
6000 Broken Sound Parkway NW, Suite 300, Boca Raton, FL 33487-2742

and by CRC Press
4 Park Square, Milton Park, Abingdon, Oxon, OX14 4RN

CRC Press is an imprint of Taylor & Francis Group, LLC

Library of Congress Cataloging-in-Publication Data (applied for)

ISBN: 978-1-032-18878-2 (hbk)
ISBN: 978-1-032-18879-9 (pbk)
ISBN: 978-1-003-25675-5 (ebk)

DOI: 10.1201/9781003256755

Typeset in Palatino
by Radiant Productions

Preface

Blockchain is assisting enterprises in digitally transforming their supply chain management activities. Supply chain management (SCM) is defined as the "systematic, strategic coordination of the traditional business functions and the tactics across these business functions within a supply chain, for the purpose of improving the long-term performance of the individual companies and the supply chain as a whole".[i] SCM aims to create an efficient and coordinated supply chain via the development of internal and external linkages. Achieving this objective is largely dependent on how effectively organizations use their technological resources and technological capabilities. While extant research has focused on how traditional technologies such as cloud computing, Radio Frequency Identification (RFID), Internet of Things (IoT), tracking technologies, and so on have impacted SCM business practices, new technological developments are emerging such as blockchain and distributed ledger technologies (DLT) that are enabling new and innovative SCM business practices. Anecdotal evidence suggests that blockchain and DLT can lead to SCM benefits such as supply chain visibility, operational efficiency, supply chain traceability, supply chain resilience, and data security. Empirical evidence pertaining to how blockchain and DLT are achieving these benefits is scarce. Furthermore, it has been proposed that blockchain and DLT could limit the impacts of global supply chain disruptions such as Covid19. This book shares lessons learned from blockchain supply chain implementations across a range of industries. Given its scope, it is primarily intended for academics, students, researchers, and, practitioners who want to learn more about how blockchain can digitally transform supply chains.

[i] Mentzer, J.T., DeWitt, W., Keebler, J.S., Min, S., Nix, N.W., Smith, C.D. and Zacharia, Z.G., 2001. Defining supply chain management. Journal of Business logistics, 22(2): 1–25.

Contents

1

A Smart Cities Blockchain Research Agenda

A Bibliometric and Network Analysis Literature Review From 2016 to September 2021

Trevor Clohessy

1. Introduction

Blockchain is a foundational technology comprising decentralised distributed digital ledgers which record the provenance of both digital and physical assets [1]. Each participant in a supply chain possesses the same digital ledger on which transactions are only approved when consensus is reached amongst all the participants [2]. The terms and conditions of these transactions are executed through smart digital contracts which are essential to the process [2]. Blockchain enables these transactions without the need for a recognised trusted central authority to overview the process. For example, blockchain's first use case was Bitcoin, a cryptocurrency which came to prominence in 2008 upon the release of its white paper by Satoshi Nakamoto [3]. Bitcoin involves the selling and purchasing of digital coins without the need for a central authority (e.g., central bank) which are used to monitor traditional fiat currencies. The disintermediation of a central trusted authority is only made possible by the cryptographic and immutable security protocols enabled by blockchain technologies. Since

Atlantic Technological University, Renmore, Galway, Ireland.
Email: trevor.clohessy@atu.ie

2008, blockchain has been adopted by a multitude of industries including the pharmaceutical industry, the diamond industry, the financial industry, the health care industry, the military industry, etc. The reasoning for adopting blockchain technologies is underpinned by its unique features which differentiate the technology from other technologies [1]. These features included access privileges, transparency, immutability, smart contracts, decentralised consensus, and distributed trust.

Blockchain is an evolving technology which has been touted as essential to the secure operation of smart city informational technology (IT) deployments [4,5]. According to [6] a smart city embodies a city where "investments in human and social capital, transport and modern IT infrastructure fuel sustainable economic growth and a high quality of life, with a wise management of natural resources through participatory government". Essentially, smart cities incorporate the use of smart technologies which enable them to become more liveable, sustainable, and responsive to the increasing economic and social requirements of its citizens [7]. For example [7], propose a smart city government cloud platform called G-Cloud which could be used to roll out smart government services across enterprises, consumers, business, and government services (Figure 1). To create this platform governments would have to invest in a technology stack which would encompass cloud computing, internet of things (IoT) technologies, sensors, real-time data analytics capturing, monitoring, and analysing capabilities, and artificial intelligence. The data generated by the smart technology stack from transactions and processes would then have to be secured. It has been argued that blockchain

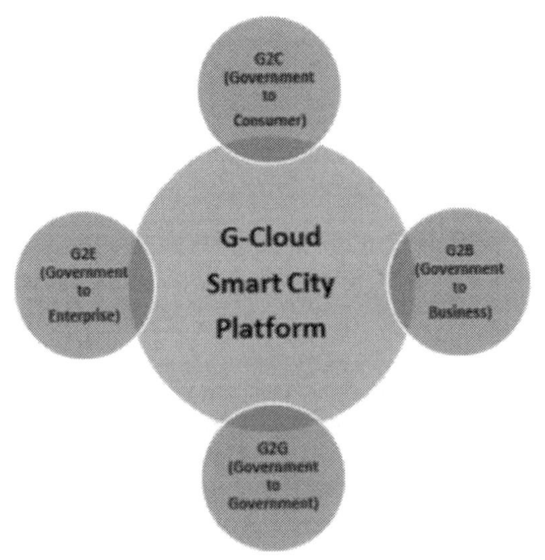

Figure 1. Smart city services platform enabled by a smart technology stack [7].

represents a technology that could excel at performing this function in terms of resiliency against security threats; scalability; fault tolerance and reliability; secure communication; and fast and efficient smart city operations [8]. Consequently, since 2016 academic research has begun to investigate how blockchain can be used in smart city contexts [9,10].

Discussions pertaining to emerging research contexts such as blockchain and smart cities can benefit from a comprehensive analysis of the extant literature in the area using bibliometric techniques to provide a holistic conceptualisation for academics and researchers. Furthermore, this mapping of the extant literature will enable researchers to identify avenues which are ripe for further investigation. Therefore, this study is centred around the following research questions (RQs):

RQ1: What are the most prominent research papers in the field of blockchain in smart cities?

RQ2: What are avenues for future research in the context of blockchain and smart city research?

The remainder of this paper is structured as follows. Section 2 provides an overview of the research methodology used to conduct the bibliometric analysis and the databases analysed. Section 3 delineates the results of the bibliometric analysis. Section 4 delineates the results of the study's citation network analysis while Section 5 presents a roadmap for future blockchain smart city research. The study concludes in section 6 with a summary of the research contributions.

2. Research Methodology

The following section details the research methodology that was executed to conduct the bibliometric analysis. While the terms "Smart City" and "Smart Cities" has a lineage dating back to the 60s [7], the terms "Blockchain", "Smart Contract", and "Distributed Ledger Technology" have only really come to prominence from an academic literature perspective since 2016. Consequently, our research focused on the time frame which spanned the years 2016–2021. For example, Figure 2 illustrates the proliferation of blockchain literature in a smart city context in the Web of Science database from January 2016 to August 2021.

2.1 Literature Review Strategy

The term "blockchain" is often used interchangeability with the words "distributed ledger technology" or "smart contracts". As a result, we extended our search criteria to include these terms. Additionally, during the initial data collection process, it was noted that the term "blockchain"

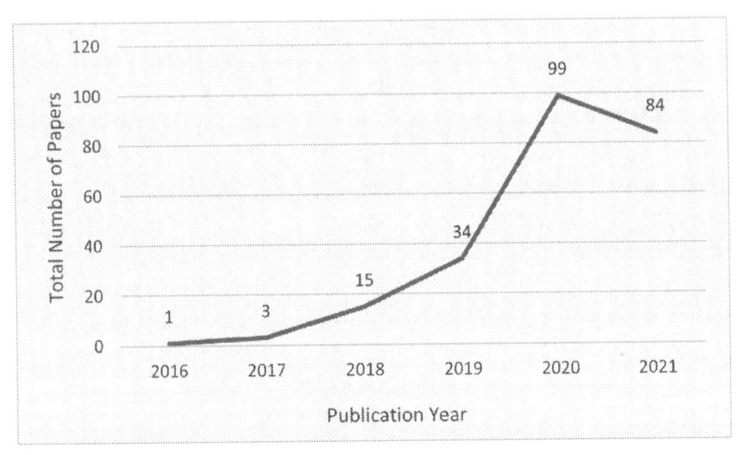

Figure 2. Smart city blockchain literature proliferation since 2016.

is sometimes spelled as "block chain". We revised our search strategy to include this keyword. Consequently, the following four different literature search combinations were used in conjunction with title, abstract, and keywords:

1. Smart City OR Smart Cities AND blockchain;
2. Smart City OR Smart Cities AND block chain;
3. Smart City OR Smart Cities AND distributed ledger technology;
4. Smart City OR Smart Cities AND smart contract

Table 1 provides an overview of the databases that were used to search for the 4 search terms combination scenarios outlined above. As can be seen, 10 databases were selected to search for these different combination scenarios rather than focusing on one specific database. This strategy is in line with previous studies [11,12] which use multiple databases for bibliometric analysis due to the concentration of blockchain research being dispersed across different academic and research domains.

2.2 Preliminary Findings

Figure 3 highlights the initial results for our defined search terms within the title, abstract and/or keywords search criteria across the 10 databases. As can be seen, Google Scholar and Web of Science contributed to 64.85% of our search results. SSRN, Emerald Insight and Taylor and Francis were relatively inefficient contributing a cumulative total of 1%. An initial Google Scholar search produced over +17,000 results. The advanced search function was used to narrow the results down to the findings depicted in Table 1.

Table 1. Total number of articles collected per database.

#	Database	Paper Number	Contribution Percentage
1	Google Scholar	306	33.81%
2	Web of Science	281	31.04%
3	EBSCO	73	8.06%
4	IEE Explore	66	7.29%
5	Springer	62	6.85%
6	MDPI	56	6.18%
7	Science Direct	51	5.63%
8	SSRN	4	0.44%
9	Emerald Insight	3	0.33%
10	Taylor and Francis	3	0.33%
	Total	905	100%

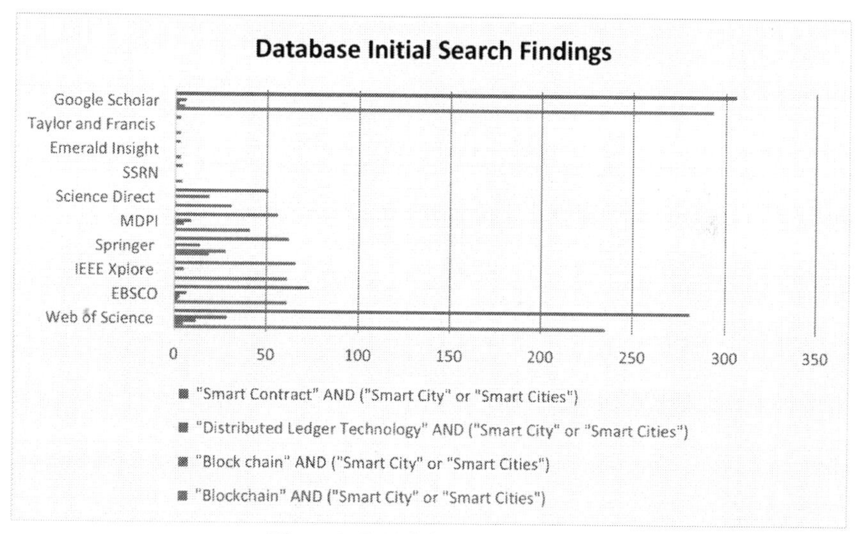

Figure 3. Initial database findings.

The purpose of this study is to review the status of blockchain and DLT smart city literature since 2016. Consequently, a structured literature review (SLR) approach, as highlighted in Figure 4, using the guidelines provided by [13], was used to ensure a high degree of replicability and transparency in the literature synthesis process. According to [14] "an SLR is a methodologically rigorous review of research results. The aim of an SLR is not just to aggregate all existing evidence on a research question; it is also intended to support the development of evidence-based guidelines

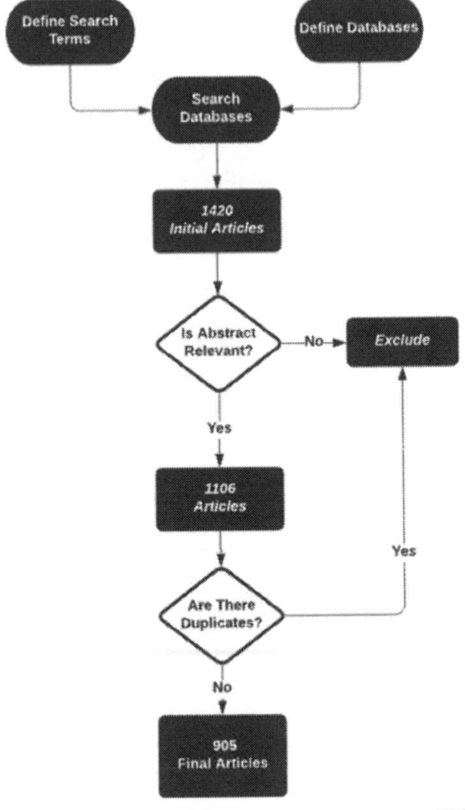

Figure 4. Structured literature review process [13].

for practitioners". Once the search terms and the databases had been defined the initial search elicited a total of 1420 articles. We did not include grey literature in terms of white papers, industry reports, theses, discussion papers and work in progress papers. The next stage of the SLR process involved the reading of abstracts of the papers. Abstracts deemed irrelevant were excluded from the next stage of analysis. For instance, the use of the word "smart city" led to non-related smart city contexts. For example, while a paper by [15] mentions blockchain in the title and abstract. The use of the word "smart city" in the abstract is not relevant to an actual "smart city" as illustrated by the following excerpt from the abstract: "The gaming industry has evolved into a multi-functional smart city that combines integrated casinos and entertainment" [15]. Ultimately, for abstracts to merit inclusion there had to be a focused connection between blockchain and DLT in a smart city context. This refinement of abstracts led to a total of 1106 abstracts. The next stage of the process led to the removal of 201 duplicate articles leaving a total of 905 articles for inclusion for this study. EndNote was used to collect and analyse the data and to remove duplicates. With regards to the latter, the "Find Duplicates" option was used in EndNote. In some instances, specific journals had recorded research papers more than once. In other cases, research papers had been categorised mistakenly (e.g., issue numbers) across journals.

3. Bibliometric Analysis Results

To answer research questions 1 and 2, a bibliometric analysis of the extant blockchain and smart city academic literature was conducted. Bibliometric approaches, often known as "Bibliometric Analysis", have become a well-established and important element of research assessment methodology, particularly in scientific and practical domains [16]. According to [17] "this type of analysis is based on the identification of the corpus of literature, i.e., publications in their broadest sense, within a given subject area". Blockchain research in a smart city context has achieved sufficient maturity to allow a bibliometric analysis of the extant research literature. A wide range of quantitative statical tools are now available to conduct bibliometric analyses. For this research, manual quantitative methods of analysis were used in conjunction with software such as Publish or Perish [18] and VOSviewer (described later). Publish or Perish is a software tool that sources and analyses scholarly citations. It obtains citations from Google Scholar, analyses them, and generates a set of citation analysis metrics. The citation results can be extracted to other software programmes or saved as a text file for future analysis. The following sections provide an overview of the results from our analysis using the aforementioned software tools.

3.1 Keyword Statistics

Researchers catalog their research articles by assigning them specific keywords. They are the terms used by researchers to describe the issue and reflect the major elements of their study topic. The keyword statistics procedure was carried out manually as part of the data gathering process to highlight the most frequently used phrases/words in the research title and keyword section. We gathered a total of 3309 keywords from our final 905 sources. Table 2 lists the most popular keyword terms. Additionally, Table 3 outlines the top ranking title keywords contained within those final set of 905 papers. Juxtaposing the two tables, it can be seen that "Blockchain/Blockchain Technology" is ranked as the number one most ranked keyword and title keyword. The second most ranked keyword is "Smart City", and the second most ranked title keyword is "Smart Cities". As these terms were used in our keyword search string

Table 2. Keyword statistics.

# Rank	Keyword	# of occurrences
1	Blockchain/Blockchain Technology	454
2	Smart City	280
3	Security	279
4	Smart Cities	273
5	IoT	209
6	Data	144
7	Privacy	106
8	Smart Contract	51
9	Distributed Ledger Technology/Distributed Ledger/DLT	39
10	Cryptocurrency	16

Table 3. Title keyword statistics.

# Rank	Keyword	# of occurrences
1	Blockchain/Blockchain Technology	748
2	Smart Cities	430
3	Smart City	309
4	IoT	308
5	Data	173
6	Security	164
7	Privacy	156
8	Sustainable	79
9	Health	68
10	Smart Contract	30

strategy their ranking was not surprising. Interestingly, the terms "Security", "IoT", "Data" and "Privacy" occupy high rankings in both tables. While it is not uncommon for these terms to be associated with blockchain technologies [1], it would appear that blockchain technologies also have a role to play in addressing smart city challenges. The keyword "Smart Contract" also appears in both tables. Smart contracts are a core characteristic of blockchain technologies. Without the need for a single administrator, blockchain allows peer-to-peer network users to share data with a high degree of dependability and transparency [1]. Smart cities have a diverse set of stakeholders, and data sharing among them is critical for providing high-quality smart urban services [7]. Smart contracts facilitate the automatic secure digital execution of the terms and conditions of

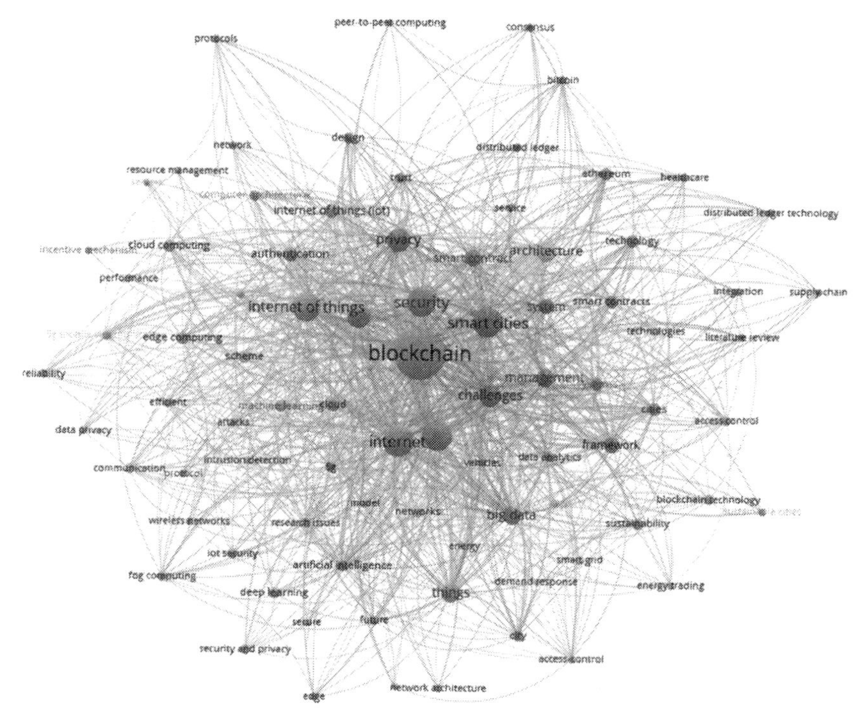

Figure 5. Keywords co-occurrence.

transactions amongst all smart city supply chain stakeholders [19]. Based on this analysis smart contracts represent an area which are ripe for further investigation in a smart city context.

4. Network Analysis of Citations

The next phase of this research incorporated a network analysis of citations for our collated research papers. A network analysis of journal citations can be used to study the various relationships between authors, journals, and research fields in a transparent and purposeful manner [20]. Furthermore, a network analysis of citations has become standard practice in research scholarship [21]. There are a number of research tools which can be used to conduct network citation analysis, and example of these include: Gelphi, CiteSpace, Pajek and VOSviewer. For this study, VOSviewer was used to analyse our bibliographic datasets. VOSviewer is a software tool which can be used for visualising and creating bibliometric networks (e.g., journals, researchers, and publications). VOSviewer enables researchers to build networks based on specific permutations including citations, co-citation, co-authorship, and bibliographic coupling. In each instance, nodes represent a published journal paper with an interconnecting line representing the citation between journal papers. The next sections delineate the various networks that were created using VOSviewer.

4.1 Citation Analysis

To investigate the network connectivity between our research journal papers/nodes, the 915 nodes were imported into VOSviewer as a text data file. Figure 6 highlights the resulting citation network. The network's initial visualization offers a sense of the richness and density of the current literary landscape. This figure is complimeneted with Table 4. Using the Publish or Perish software it was determined that the study conducted by [4] represented the most widely cited paper (largest node) with 1300+ citations.

Figure 7 represents the bibliographic coupling network (BCN) that was created using VOSViewer. According to [22] a BCN represents "shared references among publications, thus it provides deeper insights on the scientific activity, as it reveals information on how authors use and construct links among the existing literature." The authors argue that BCN can be used to examine a particular research study's stance in the field to determine the size of its research community, and it may assist in solving

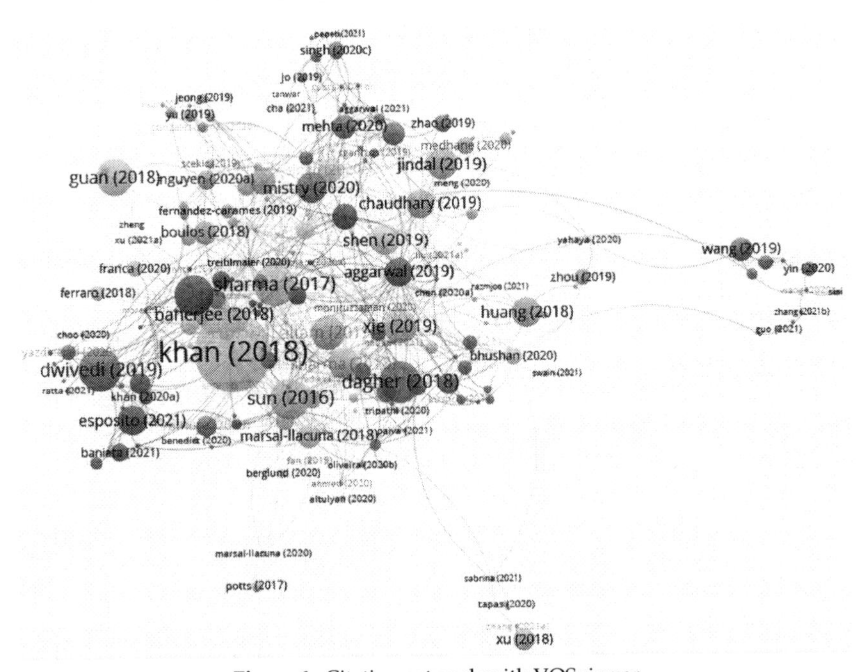

Figure 6. Citation network with VOSviewer.

Table 4. Citation ranking.

Citation Rank	Author	#Citations
1	[4]	1349
2	[8]	492
3	[41]	407
4	[25]	371
5	[42]	353
6	[43]	352
7	[44]	343
8	[45]	335
9	[46]	327
10	[47]	307

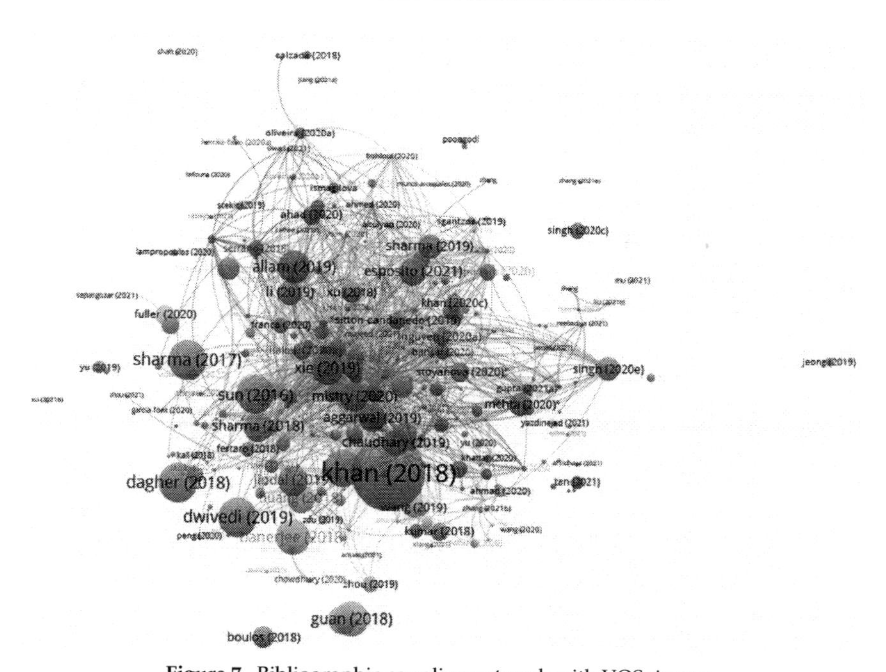

Figure 7. Bibliographic coupling network with VOSviewer.

unanswered questions. Ultimately, BCN analysis determines the degree of overlap in the bibliography lists of publications.

4.2 Co-Citation Analysis

Figure 8 depicts a co-citation network visualization for the dataset used in our investigation. The co-occurrence of journal articles in a research paper's reference list is represented by co-citation data analysis, which is an exploratory kind of bibliographic analysis [23]. Research articles are co-cited if they appear in the reference list of another research paper, according to [24]. If research articles A and B exist in the reference list of research paper C, they can be referenced together. Research articles having a large number of co-citations are more likely to be connected and in the same field(s). As can be seen from figure 6 the papers by [25] and by [3] are the two most widely cited papers contained with the reference lists of blockchain and smart city research studies. In the context of [3] this is understandable as Bitcoin represents the first use of blockchain technology and this 9 page whitepaper represents the first introduction of the term blockchain. The whitepaper also provides an overview of how blockchain technology functions. With regards to the [25] paper, this research study focuses on how blockchain can be used in a smarty city context for intelligent transport systems. The study proposes a distinct smart city intelligent vehicle network architecture underpinned by blockchain technology called Block-VN.

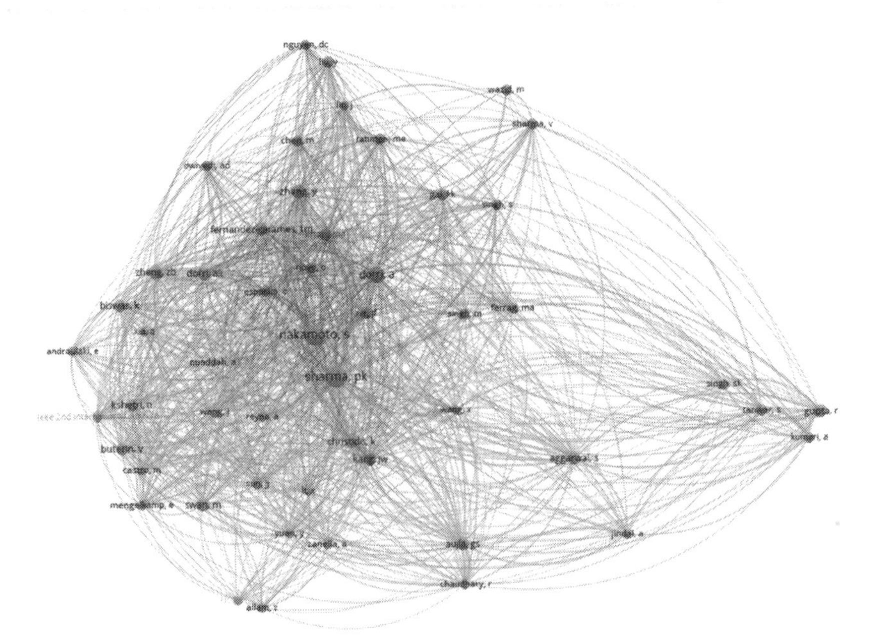

Figure 8. Co-citation network with VOSviewer.

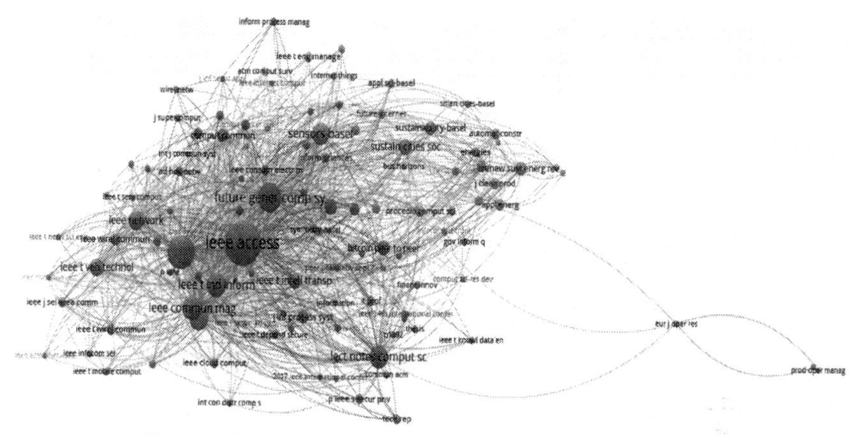

Figure 9. Journal co-citation sources network with VOSviewer.

In addition to the authors, another important co-citation analysis factor are journal sources. For the purposes of this study, a co-citation analysis was conducted of the main journals sources in which the authors were co-cited. This form of analysis can provide perspective pertaining to the intellectual base of a specific knowledge domain. Consequently, the journal sources network depicted in Figure 9 contains just over 110 journals sources with equal or more to 25 citations. The top journal sources in terms of co-citation frequency are IEEE Access, Future Generation Computer Systems, IEEE Computer Magazine, Lecture Notes in Computer Science and Sensors (Open Access Journal from MDPI). These results reveal software engineering and computer based nature of extant research on blockchain and smart city research. There is a need for interdisciplinary research in the field of study, particular from business and information systems perspectives. For example, the information systems discipline denotes the senior basket of 8 journals as top in their field [26]. A review of these basket journals indicated that there is a dearth of research which has focused on how blockchain can be applied in a smart city context.

5. Smart City Blockchain Research Agenda

Our study has highlighted how blockchain research has been explored in various smart city contexts from IoT, security, privacy, mobility, energy, and public services. In terms of research question 2, Figure 10 illustrates the various research categories which this study has identified as areas for future research based on our bibliometric and co-citation network analysis of the blockchain and smart city research. We will now discuss these future research areas in turn.

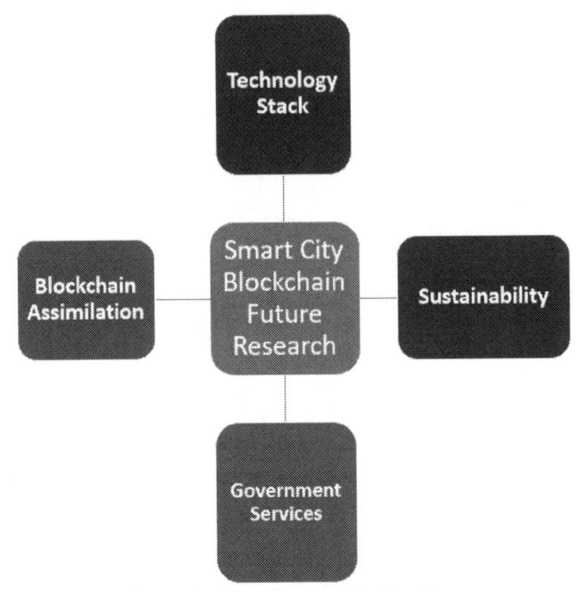

Figure 10. Future research directions.

Research Direction #1: Government Services

From a citizen perspective, blockchain and distributed ledger technologies have the potential for global governments to roll out citizen-centric services which are reliable, secure, private, and trustful [27,28]. Blockchain can overcome existing paper laden processes which are time consuming, costly, and inefficient. Academic degrees, occupational licenses, employee IDs, and other identification credentials are easier to deal with when they're digital. However, this extremely personal data must be kept hidden and safe. There is now empirical evidence and ongoing global use cases to highlight how blockchain is being used by governments, corporations, and educational institutions to allow a secure and trustworthy infrastructure and improve services. Covid19 has accelerated governmental digital transformation. Take for instance the introduction of Covid19 tracking and tracing applications and digital vaccination certificates. There is now an opportunity to investigate how blockchain can be used to secure government centric-services as highlighted previously in Figure 1. From an industry perspective, blockchain possesses the potential to transform how governments formulate strategies with regards to smart health, smart energy, smart mobility, smart voting, smart education, and smart logistics. For example, from a healthcare perspective, blockchain and DLT can assist healthcare practitioners facilitate: the safe transfer of patient medical records, the managing of drug supply chain logistics, and the secure storage and transfer of patient medical records. From a voting perspective blockchain technologies may be used to overcome concerns with traditional electronic voting, making it more cost-effective, reliable, and secure [29]. Additionally, from a smart mobility perspective, blockchain technology can be used for exchanging data for automobiles, such as ride-sharing, car sharing, or self-driving cars [30]. Research exploring how blockchain can assist governments across various smart public service sectors merits further scrutiny and debate by researchers in terms of the possible negative consequences of blockchain with regards to the immutability of citizen records and the potential for privacy violations [31].

Research Direction #2: Sustainability

Smart city definitions vary but two concepts which remain core to each definition is the concept of sustainability [32] and how technology can be used to enhance sustainability from socio-technical and socio-economic perspectives (Bosch, 2019). For example, [33] define a smart city as "a loosely connected set of confluences between data, digital technologies, and urban sites and processes", which can be catalysed by a "digitally enabled data-driven, continually sensed, responsive and integrated urban environment". There are ongoing global efforts to explore how blockchain can facilitate smart city sustainability agendas. Future research can investigate how smart cities can use blockchain to develop their sustainability strategies in line with global climate change and sustainability action plans. For example, research which investigates how smart cities can construct secure and shared digital ledgers for real-time data management in transportation, energy, and utilities would be welcome. Ultimately, blockchain technology's application might assist smart cities in streamlining their interactions with citizens, reducing individual and collective resource usage, and sharing public data with approved third parties. Blockchain can also assist citizens to reduce, metricise and analyse their individual carbon footprints and assist with the dematerialization of industrial supply chains [34].

Research Direction #3: Blockchain Assimilation

While this research has identified that blockchain smart city research is increasing, it has also identified that the research area is at an embryonic stage. Consequently, further research is required to identify successful blockchain smart city research case study success stories and research which also examine case study failures. These research studies can be guided using the concept of assimilation. Assimilation can be defined as "an organisational process that "(i) is set in motion when individual organisation members first hear of an innovation's development, (ii) can lead to the acquisition of the innovation, and (iii) sometimes comes to fruition in the innovation's full acceptance, utilization, and institutionalization" [35]. The assimilation of a technological innovation can often be challenging and is rarely binary [36]. Assimilation theory posits that assimilation may intensify or deteriorate over the course of a technological innovation's adoption journey [37]. Each assimilation stage describes the degree to which the technological innovation permeates the adopting company. Often the causes of innovation success or failures can be minute. While many frameworks have been proposed to understand how innovation assimilation impact entities [38], one of the most widely cited and used is proposed [36]. Table 5 provides an overview of this six-stage model. This framework comprises two pre-adoption 'early stages' (e.g., initiation and adoption) and 4 post adoption 'later stages' (e.g., adaptation, acceptance, routinisation, and infusion). The infusion stage comprises several different facets of technology innovation infusion. When investigating assimilation stages, a salient consideration is the degree to which we may expect an adopting organisation's progression

Table 5. Innovation assimilation stages [36].

Stage	Description
Initiation	A match is found for a technological innovation and its application in an organisation.
Adoption	A decision is reached to invest resources to adopt the technological innovation.
Adaptation	The technological innovation is developed, installed, and maintained. Organisational members are trained to use the new technological innovation.
Acceptance	Organisational members are induced to use the technological innovation which is now being used within the company.
Routinisation	The use of the technological innovation is encouraged as a normal activity and organisational structures are altered to accommodate the technological innovation. It is no longer seen as something out of the ordinary.
Infusion	The technological innovation is used in an elaborate and sophisticated manner. Infusion is categorised in several ways: • **Extended use:** using more features of the technological innovation; • **Integrative use:** using the technological innovation to create new workflow linkages among tasks; • **Emergent use:** using the technological innovation to perform tasks not previously considered possible.

through the assimilation stages to be linear. However, extant research has demonstrated that progression may be non-linear. For instance, [37] argue that a sequential model is more likely to emerge for 'off-the-shelf' technologies in comparison to 'bespoke' technological innovations. Further, [37] suggest that an unfreeze (e.g., initiation, adoption, stages) and unfreeze (e.g., acceptance, routinisation, and infusion stages) sequential pattern is likely to emerge.

This six-stage assimilation framework has been used in previous studies that examine technological adoption [39]. We also believe it would also be an appropriate lens with which to examine the benefits and challenges associated with the adoption of blockchain technology in smart city contexts. Of particular interest would be future research relating to the infusion assimilation stage which could be sued to identify the "emergent use" cases for blockchain in smart city contexts.

Research Direction #4: Technology Stack

This study identified that blockchain is not used in isolation in smart city contexts. The technology is used in fusion with other technologies such as cloud computing, artificial intelligence, fog computing, sensor based IoT, drone technologies, ubiquitous technologies and so on. This study argues that future blockchain research should focus on how blockchain technologies can be deployed in smart city contexts using a concept we refer to as the technology stack. A technology stack encompasses the overall technology infrastructure and data ecosystem that is constructed when deploying smart technology applications. This study argues that blockchain technology and DLT can serve as the foundational layer which provides security, immutability, privacy and trust for the technologies and data which are built/deployed on it. For example, [40] explored how DLT can be deployed in a medical supply chain using IoT sensor, business intelligence and cloud based technologies to create vigilant information based systems. This multi-layer technology stack can be used to enable organisations to make quick decisions in real-time in dynamic supply chain environments. It can also assist with the prevention of counterfeit medications. Furthermore, "the application of smart technology stacks which are underpinned by blockchain and DLT can equip smart city decision-makers with the data need for the planning, development and maintenance of smart city services, resources, and infrastructure in real-time.

6. Conclusion

This study set out to answer the following two research questions:

RQ1: What are the most prominent research papers in the field of blockchain in smart cities?

RQ2: What are avenues for future research in the context of blockchain and smart city research?

By investigating these two research questions this study has achieved the following theoretical contributions. In terms of research question 1, the study's bibliometric analysis and citation network analysis methodology revealed not only the most prominent papers in smart city blockchain research but also revealed that most of the extant research is inherently software engineering and computer based. Consequently, opportunities exist in other research disciplines to explore how blockchain can applied in a smart city context. Ultimately, further interdisciplinary research is required to foster greater critical thinking and to synthesis concepts from other disciplines to advance smart city blockchain research. With regards to research question 2, the study identified four distinct categories of research directions which merit further investigation which can assist researchers to identify further workable blockchain smart city solutions. Ultimately, the ongoing rapid digitalization of global economies means that governments must now consider how blockchain and DLT can be applied in their smart strategies. This study has highlighted that while research in the area is at an embryonic stage, there is sufficient empirical evidence that blockchain and DLT can be effective in ensuring smart and sustainable urban development. While the current study provides and in-depth analysis of the blockchain smart city literature from 2016 to 2021, we encourage other researchers to also explore the area via bibliometric and citation network analysis techniques.

References

[1] Clohessy, T. and Acton, T. 2019. Investigating the influence of organizational factors on blockchain adoption: An innovation theory perspective. *Industrial Management & Data Systems*.

[2] Sigalov, K., Ye, X., König, M., Hagedorn, P., Blum, F., Severin, B., Hettmer, M., Hückinghaus, P., Wölkerling, J. and Groß, D. 2021. Automated Payment and Contract Management in the Construction Industry by Integrating Building Information Modeling and Blockchain-Based Smart Contracts. *Applied Sciences*, 11(16): 7653.

[3] Nakamoto, S. 2008. Bitcoin: A peer-to-peer electronic cash system. *Decentralized Business Review*, 21260.

[4] Khan, M. A. and Salah, K. 2018. IoT security: Review, blockchain solutions, and open challenges. *Future Generation Computer Systems*, 82: 395–411.

[5] Sabrina, F. and Jang-Jaccard, J. 2021. Entitlement-based access control for smart cities using blockchain. *Sensors*, 21(16): 5264.

[6] Caragliu, A., Del Bo, C. and Nijkamp, P. 2011. Smart cities in Europe. *Journal of Urban Technology*, 18(2): 65–82.

[7] Clohessy, T., Acton, T. and Morgan, L. 2014. Smart city as a service (SCaaS): A future roadmap for e-government smart city cloud computing initiatives. In *2014 IEEE/ACM 7th International Conference on Utility and Cloud Computing* (pp. 836–841). IEEE.

[8] Biswas, K. and Muthukkumarasamy, V. 2016. Securing smart cities using blockchain technology. In *2016 IEEE 18th international conference on high performance computing and communications; IEEE 14th international conference on smart city; IEEE 2nd international conference on data science and systems (HPCC/SmartCity/DSS)* (pp. 1392–1393). IEEE.

[9] Botello, J. V., Mesa, A. P., Rodríguez, F. A., Díaz-López, D., Nespoli, P. and Mármol, F. G. 2020. BlockSIEM: Protecting smart city services through a blockchain-based and distributed SIEM. *Sensors*, 20(16): 4636.

[10] Vintimilla-Tapia, P., Bravo-Torres, J., López-Nores, M., Gallegos-Segovia, P., Ordóñez-Morales, E. and Ramos-Cabrer, M. 2020. VaNetChain: a framework for trustworthy exchanges of information in VANETs based on Blockchain and a virtualization layer. *Applied Sciences*, 10(21): 7930.

[11] Kousis, A. and Tjortjis, C. 2021. Data Mining Algorithms for Smart Cities: A Bibliometric Analysis. *Algorithms*, 14(8): 242.

[12] Müßigmann, B., von der Gracht, H. and Hartmann, E. 2020. Blockchain technology in logistics and supply chain management—A bibliometric literature review from 2016 to January 2020. *IEEE Transactions on Engineering Management*, 67(4): 988–1007.

[13] Kitchenham, B. 2004. Procedures for performing systematic reviews. *Keele, UK, Keele University*, 33: 1–26.

[14] Kitchenham, B., Brereton, O. P., Budgen, D., Turner, M., Bailey, J. and Linkman, S. 2009. Systematic literature reviews in software engineering–a systematic literature review. *Information and Software Technology*, 51(1): 7–15.

[15] Liao, D. Y. and Wang, X. 2018. Applications of blockchain technology to logistics management in integrated casinos and entertainment. In *Informatics* (Vol. 5, No. 4, p. 44). Multidisciplinary Digital Publishing Institute.

[16] Wallin, J. A. 2005. Bibliometric methods: Pitfalls and possibilities. *Basic and Clinical Pharmacology and Toxicology*, 97(5): 261–275.

[17] Ellegaard, O. and Wallin, J. A. 2015. The bibliometric analysis of scholarly production: How great is the impact?. *Scientometrics*, 105(3): 1809–1831.

[18] Harzing 2021. Publish or Persih. [Online]. Available: https://harzing.com/pophelp/.

[19] Palaiokrassas, G., Skoufis, P., Voutyras, O., Kawasaki, T., Gallissot, M., Azzabi, R., Tsuge, A., Litke, A., Okoshi, T., Nakazawa, J. and Varvarigou, T. 2021. Combining blockchains, smart contracts, and complex sensors management platform for hyper-connected smartcities: An iot data marketplace use case. *Computers*, 10(10): 133.

[20] Freeman, I. 2004. The developmental of social network analysis: A study of the sociology of science. Vancouver: Empirical Press.

[21] Barnett, G. A., Huh, C., Kim, Y. and Park, H. W. 2011. Citations among communication journals and other disciplines: a network analysis. *Scientometrics*, 88(2): 449–469.

[22] Biscaro, C. and Giupponi, C. 2014. Co-authorship and bibliographic coupling network effects on citations. *PloS one*, 9(6): e99502.

[23] Ding, Y., Yan, E., Frazho, A. and Caverlee, J. 2009. PageRank for ranking authors in co-citation networks. *Journal of the American Society for Information Science and Technology*, 60(11), 2229–2243.

[24] Chang, C. Y., Gau, M. L., Tang, K. Y. and Hwang, G. J. 2020. Directions of the 100 most cited nursing student education research: A bibliometric and co-citation network analysis. *Nurse Education Today*, 104645.

[25] Sharma, P. K., Moon, S. Y. and Park, J. H. 2017. Block-VN: A distributed blockchain based vehicular network architecture in smart city. *Journal of Information Processing Systems*, *13*(1): 184–195.

[26] Association for Information Systems. Senior Scholars' Basket of Journals. Retrieved at: https://aisnet.org/page/SeniorScholarBasket on June 8th 2021.

[27] Bhushan, B., Khamparia, A., Sagayam, K. M., Sharma, S. K., Ahad, M. A. and Debnath, N. C. 2020. Blockchain for smart cities: A review of architectures, integration trends and future research directions. *Sustainable Cities and Society*, 61: 102360.

[28] Rivera, R., Robledo, J. G., Larios, V. M. and Avalos, J. M. 2017. How digital identity on blockchain can contribute in a smart city environment. In *2017 International smart cities conference (ISC2)* (pp. 1–4). IEEE.

[29] Jafar, U., Aziz, M. J. A. and Shukur, Z. 2021. Blockchain for electronic voting system—review and open research challenges. *Sensors*, 21(17): 5874.

[30] Berneis, Moritz, Devis Bartsch and Herwig Winkler. "Applications of Blockchain Technology in Logistics and Supply Chain Management—Insights from a Systematic Literature Review." *Logistics* 5.3 (2021): 43.

[31] Picone, Marco, Cirani Simone and Veltri Luca. 2021. Blockchain Security and Privacy for the Internet of Things. *Sensors* 21(3): 1–4.

[32] Toli, A.M. and Murtagh, N. 2020. The Concept of Sustainability in Smart City Definitions. *Front. Built Environ.* 6: 77. doi: 10.3389/fbuil.2020.00077.

[33] McFarlane, C. and Söderström, O. 2017. On alternative smart cities: from a technology-intensive to a knowledge-intensive smart urbanism. *City* 21: 312–328. doi: 10.1080/13604813.2017.1327166.

[34] Chen, J., Cai, T., He, W., Chen, L., Zhao, G., Zou, W. and Guo, L. 2020. A blockchain-driven supply chain finance application for auto retail industry. *Entropy*, 22(1): 95.

[35] Meyer, A.D. and Goes, J.B. 1988. Organisational assimilation of innovation: a multilevel contextual analysis, Academy of Management Journal 31: 897–923.

[36] Gallivan, M. 2001. Organisational adoption and assimilation of complex technological innovations: development and application of a new framework. *Database for Advances in Information Systems*, 32: 51–85.

[37] Cooper, R.B. and Zmud, R.W. 1990. Information technology implementation research: a technological diffusion approach. *Management Science*, 26: 123–39.

[38] Saga, V. L. and R. W. Zmud. 1994. The nature and determinants of IT acceptance, routinization, and infusion. Levine L., ed. Diffusion, Transfer and Implementation of Information Technology, 67–86.

[39] Basole, R. C. and Nowak, M. 2018. Assimilation of tracking technology in the supply chain. *Transportation Research Part E: Logistics and Transportation Review*, 114: 350–370.

[40] Clohessy, T. and Clohessy, S. 2020. What's in the Box? Combating Counterfeit Medications in Pharmaceutical Supply Chains with Blockchain Vigilant Information Systems. In *Blockchain and distributed ledger technology use cases* (pp. 51–68). Springer, Cham.

[41] Sun, J., Yan, J. and Zhang, K. Z. 2016. Blockchain-based sharing services: What blockchain technology can contribute to smart cities. *Financial Innovation*, 2(1): 1–9.

[42] Dagher, G. G., Mohler, J., Milojkovic, M. and Marella, P. B. 2018. Ancile: Privacy-preserving framework for access control and interoperability of electronic health records using blockchain technology. *Sustainable Cities and Society*, 39: 283–297.

[43] Zhao, J. L., Fan, S. and Yan, J. 2016. Overview of business innovations and research opportunities in blockchain and introduction to the special issue. *Financial Innovation*, 2(1):1-7.

[44] Talari, Saber, Miadreza Shafie-Khah, Pierluigi Siano, Vincenzo Loia, Aurelio Tommasetti, and João PS Catalão. "A review of smart cities based on the internet of things concept." Energies 10, no. 4 (2017): 421.

[45] Dwivedi, A. D., Srivastava, G., Dhar, S. and Singh, R. 2019. A decentralized privacy-preserving healthcare blockchain for IoT. *Sensors*, 19(2): 326.
[46] Banerjee, M., Lee, J. and Choo, K. K. R. 2018. A blockchain future for internet of things security: a position paper. *Digital Communications and Networks*, 4(3): 149–160.
[47] Gharaibeh, A., Salahuddin, M. A., Hussini, S. J., Khreishah, A., Khalil, I., Guizani, M. and Al-Fuqaha, A. 2017. Smart cities: A survey on data management, security, and enabling technologies. *IEEE Communications Surveys & Tutorials*, 19(4): 2456–2501.
[48] Bosch. *Smart City Concepts–The City of Tomorrow*. [Online]. Retrieved at: https://www.bosch.com/stories/smart-city-challenges/pdf on May 5th 2020.

2

Breaking Bitcoin at any Cost

Irrational Attacks on Proof-of-Work Blockchain

Ashish Rajendra Sai

1. Introduction

The unrest caused by the 2009 economic crisis is thought to have influenced the launch of Bitcoin, the first decentralized digital currency [24]. Bitcoin proposes a method of value transfer without the use of a trusted third party in an untrustworthy environment [12]. Bitcoin accomplishes this by employing a peer-to-peer network that maintains a global ledger of all transactions. A consensus algorithm is used to reach an agreement on the ledger's view of the data, in which the majority of network participants agree on a single view of the data.

The novel contribution of Bitcoin in the field of digital currencies is the application of a clever incentive mechanism for participants of the network. The participants of the Bitcoin network are anticipated to solve a computationally expensive trivial mathematical problem. The first participant to solve the mathematical problem acquires a reward for the computational cycles spent on the solution. This numerical solution is frequently referred to as Proof-of-Work. The reward mechanism serves as an incentive for honest behavior in the network. Dishonest behavior is de-incentivized in the form of lost rewards if half of the network dissents with the attackers' view of the data. It is presupposed based on the economic barriers that half of the network participants are honest.

University of Amsterdam.
Email: a.r.k.sai@uva.nl

Bitcoin's underlying technology Blockchain has seen numerous applications including supply chain. The utility of Blockchain in the supply chain relies on the secure operation of the underlying consensus mechanism.

Because the financial asset is entangled in the consensus, it is assumed that an attacker would try to maximize the profit from the attack. This assumption of a rational miner is employed by numerous Bitcoin security evaluation frameworks [32], [15], [35], [18], [14], [41], [31]. We believe that the assumption may limit the applicability of current attacking strategies that do not account for extreme scenarios in which the adversary is unconcerned about any financial gain or loss incurred during the attack.

In this chapter, we aim to model an irrational attacker. We describe the irrationality as the intent of harming the network, notwithstanding the cost (remunerative or reputational). The chapter intends to answer the following research question:

What attack strategies could an irrational attacker use to put the bitcoin network in jeopardy?

To answer the research questions, we devise four dimensions that an irrational attack can deploy attack vectors on to exploit the network. The design of Bitcoin determines the choice of these dimensions. The bitcoin core protocol is an implementation guideline for the participants of the network. These guidelines are realized in the core client, which is deployed over a peer-to-peer network of participants. We describe the implementation of guideline oriented attack vectors as Protocol Layer Attacks. The protocol by its design induces a race condition for the proof-of-work in the network to establish consensus on the view of the data.

The attack vectors associated with the race conditions are defined as Consensus Layer Attacks. The consensus pivots copiously on the transmission of information between the participants of the network. The attack vectors that strive to exploit the transmission of information are categorized as Network Layer Attacks.

The classifications listed above are imperative for accurate computational execution of the implementation guidelines, but one riveting factor that affects the network is the economics involved in the incentives for the participation. We categorize attacks on the incentives as Economic Layer Attacks. We can recapitulate the attacks modelled in this chapter in four categories: Protocol, Consensus, Network, and Economic layer attacks. The chapter makes the following contributions:

1. A novel investigation of attack vectors for an irrational attacker on the Bitcoin network (Section 3).

2. The chapter proposes new attack strategies on the Protocol layer including the Nonce Attack (Section 4.1) and Difficulty Recalibration Attack (Section 4.1).
3. We propose and appraise the possibility of an Extortion Attack on the Consensus Layer of the network (Section 4.2).
4. We introduce novel Denial of Service attack vectors for an Irrational Attacker on the Network Layer of Bitcoin (Section 4.3).
5. Provided a wealthy attacker; we model the feasibility of an Economic attack on the Bitcoin network by modelling artificially fabricated panic selling and a novel Griefer attack (Section 4.4).

2. Background

In this section, we first probe the structure of a blockchain followed by how Bitcoin implements a blockchain data-structure on peer-to-peer distributed systems. We also list the prominent attacks on the Proof-of-Work (PoW) based blockchains.

2.1 Structure of Bitcoin

The term Blockchain is often used as an umbrella term to refer to the broad field of Distributed Ledger Technology [33]. The term Blockchain was first used by the creator of Bitcoin Satoshi Nakamoto in a GitHub commit to note the data structure used by Bitcoin [49]. Bitcoin uses a cryptographically linked structure of blocks that accommodate transactions. Each new block of the transaction is connected to the previous block cryptographically developing a chain of blocks. The participants of the bitcoin network are tasked with the accordance on a single view of this append-only structure. To accomplish this consistency of data in an unconstrained distributed environment, Bitcoin utilizes a peer-to-peer distributed system with cleaver Proof-of-Work based incentive mechanism. The implementational details of such a system are beyond the scope of this chapter.

We cohere to the structure proposed by [25], in which the authors identify three abstract technical components to describe the complex functioning of the bitcoin network. These components are the consensus protocol, the communication network, and the transactions (including scripts). The consensus protocol is used to elect a leader via a race over the solution of a mathematically hard computational problem. The communication network component enables the participants (also referred to as Nodes) to exchange messages concerning new transactions or blocks on the network amongst other protocol level messages. The third component identified by [25] is the transactions component; in this component, the authors accumulate the protocol specifications correlated with the transactions on the bitcoin networking, including the script execution.

Even though the proposed abstracted categorization provides [25] an insightful overview of the functioning of the bitcoin network, we extend the categorization with minor alterations in the study. We design a layered architecture for the technical components of the system. The lowest layer in the proposed architecture is Protocol Layer. In the protocol layer, we define the structure of the system, including how transactions are stored and processed encompassing the transactions component proposed by [25].

Other protocol layer components include the blockchain structure used by the Bitcoin network along with the cryptographic primitives employed on the protocol level. The network participants must adhere to the protocol layer specifications to participate in the network. This adherence is often accomplished by deploying the Bitcoin-core client on the nodes.

The protocol layer specifies the behavior of the network over a distributed network of nodes (network participants). This network connection establishment and intercommunication between nodes are captured in Network Layer of the architecture. The network layer is responsible for the discovery of nodes that deploy the adhering protocol client. The network layer is also responsible for efficient communication between the nodes present in the network. The Network Layer serves as the information dissemination mechanism of the system.

The main aim of the Bitcoin network is to deterministically agree on a single view of the data under certain assumptions (the primary assumption being the requirement of the majority of nodes being honest) [47]. We describe the consensus Layer that ensures that the network reaches a consensus with some degree of assurance. Agreement between nodes over the view of data is attained through a cleaver Proof-of-Work incentive mechanism. This incentive mechanism is a part of the consensus layer in the proposed architecture.

As the article endeavors to analyze the behavior of an irrational attack, we extend the proposed technical architecture to embody the economic aspects of the system additionally. We posit that as the Bitcoin network inherently depends on the strength of the incentives to ensure that the majority of the participants are honest, it is imperative to capture the economics while analyzing the attacks. Thus we suggest a fourth layer in the layered architecture, the Economic Layer, which incorporates the economic aspects of the Bitcoin System.

2.2 Security Attacks on Bitcoin

Conventionally, payment systems have relied on a central authority to guarantee that the system is secure by validating all the transactions processed by the system. Unlike the conventional systems, bitcoin does not have a central authority, but the majority of the network is tasked with the honest validation of the transactions in the network [46]. The distributed and decentralized nature of Bitcoin introduces new attack vectors that are not present in the traditional payment systems. In this subsection, we analyze the most prominently researched security attacks on the Bitcoin network.

2.2.1 Double Spending Attack

The paramount impediment in the development of a decentralized payment system before bitcoin was the plight of a Double Spend. A double spend is defined as the payment originated with the same currency unit twice, i.e., using the same coin to pay for two transactions. Traditionally, this is solved by the central authority which maintains a ledger and can verify against the ledger if the coin has already been spent. Bitcoin provides a probabilistic guarantee that a double spending attack will not succeed with the assumption that the majority of the network is honest. One way of achieving a double spend on the Bitcoin network is by engaging a Finney Attack [55].

2.2.2 Finney Attack

An attacker who intends on double-spending transaction t_d will send the subjected transaction to a recipient simultaneously the attacker creates a conflicting transaction with a different recipient but keeps this conflicting transaction in a private chain. Once the transaction t_d is appended to a block in the Blockchain, the recipient releases the service or product to the attacker. At this point, the attacker decides to publish the block with the conflicting t_d to the Bitcoin network resulting in a fork. If the majority of the participants decide to adopt the conflicting block, the attacker has successfully spent the same bitcoin twice. The prospect of such an attack can be minimized substantially by the recipient if the recipient decides to wait for a reasonable number of block confirmation (k). The de-facto standard value of k for the Bitcoin network is considered to be 6.

In a Finney attack, the attacker exploits a private blockchain to execute the attack. These types of attacks are identified as Block Withholding Attacks [35]. Supplementary sophisticated attacks have been proposed that utilize a similar strategy of performing the mathematically intense Proof-of Work operations (also referred to as mining) in private.

2.2.3 Selfish Mining

In Selfish mining [41], the attacker aims to augment an unfair share of reward by causing harm to the other participants of the network. The attacker achieves this by mining on a private chain and publishing the private chain strategically to ensure the loss of the reward for other participants. The attacker does this with the anticipation that other participants will join the attackers' coalition.

The attackers' success probability is profoundly reliant on the direct network connection that the attacker has [41]. It is reported that network connectivity may be a decisive factor. Other attacks that exploit the network connectivity include the Eclipse Attack [34].

2.2.4 Eclipse Attack

In an Eclipse mining [26], the attacker partitions the victim from the rest of the network. The view of the data that the victim sees is a monopolized version of the actual data based on the will of the attacker. The attacker in this attack manipulates all the incoming and outgoing network connections of the victim. The attacker can exploit the eclipse miners computational power for their benefit.

These types of attacks aim to accomplish monetary gain by economically harming other entities (e.g., lost block reward). The economic attacks on the bitcoin network may prove to be most likely as the reliance on monetary value for security is a significant limitation. Some other attacks that exploit the economic properties of bitcoin include Whale Transaction [38] to reduce the growth rate of the network by promoting more forks.

2.2.5 Whale Transaction

A Whale transaction is defined as a transaction with very high monetary value as the transaction fee, which may promote forks in the network [38], [48]. All the miners are incentivized to include the whale transaction in their block to receive the significantly high transaction fee in their block. This race to include the whale transaction in the block limits the chain growth rate as most miners are indulged in the race to win the transaction fee reward.

In this attack, the authors introduced the idea of miners favoring a chain fork based on incentives. Another attack that exploits the same economic gain incentive is a Bribery attack [28].

2.2.6 Bribery Attack

In a Bribery attack [28], the attacker intends to double spend a transaction t_d by incentivizing other miners to favor the fork that includes the conflicting transaction. Combining a whale transaction to favor the fork that includes a conflicting transaction may allow for a successful double spend attack.

The attacks listed above are some of the most prominently reported attacks on the Bitcoin network in the literature. These attacks assume that the attacker seeks to attain a monetary gain by performing the attack. This assumption may seem reasonable as the cost involved in the successful execution of any of these attacks is high. However, we argue that if a largescale organization is motivated to compromise the network, they may not be incentivized by the possible gain from the attack. In this chapter, we identify more attack vectors in the architectural layers identified above for an irrational attacker. The following section provides an overview of the study design.

3. Study Design

As the Bitcoins network is appreciably intricate, an exhaustive search of all possible attack strategies is considered injudicious. We abstain from the exhaustive search by conducting a review of the architectural layers (Section 2) from the attackers' perspective. The review results in a list of potential attacks. We adhere to the reflective action science research method [3]. We do not validate if the attacks are realistic, practical or feasible, we leave this up to future research.

The reflective action science approach details a five-phase cyclic process [7], [1] consisting of Diagnosis, Action planning, Action taking, Evaluating, and Specifying Learning. In the following section, we will review the action plan in detail.

Prior to the initialization of reflective action research, a research environment must be established. In our case, the research environment is the aforementioned four-layered architecture. After the establishment of the research environment, the cyclic process is initiated with the diagnosis of the problem with the information system.

In the diagnosis phase, we aim to identify the possible attack strategies for the irrational attacker. In line with the current research in action science, the diagnosis entails self-interpretation of the problem, not by reduction and simplification, but preferably in a holistic fashion [7]. The diagnosis also assists us in the development of certain theoretical assumptions about the characteristics of the problem [9].

The diagnosis is followed by action planning phase in which researchers and practitioners collaborate to determine the methods of validation of previously identified problems. For this study, the action planning phase involves the identification of evaluation techniques for the identified problems. Theoretical frameworks traditionally drive these methods.

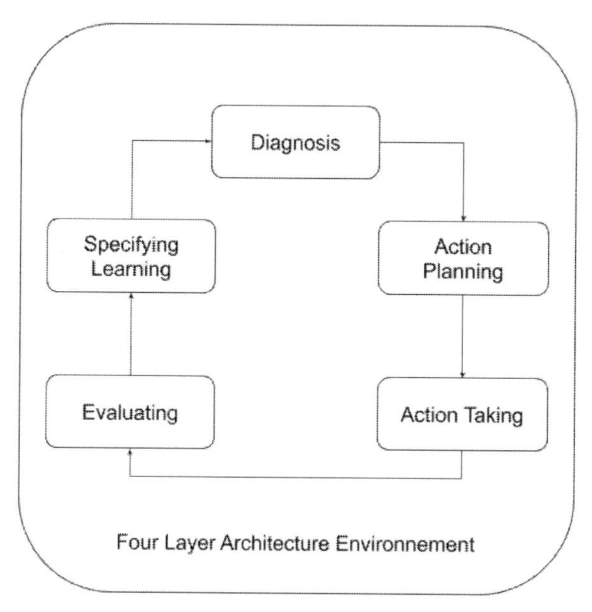

Figure 1. Action research plan.

In the next phase known as Action Taking, the identified methods of validations are implemented. After the action implementation phase, the results are evaluated by the researchers and practitioners in the Evaluation phase. The evaluation phase is followed by the Specifying Learning phase in which the knowledge gained by the action research is used to cultivate the attack strategy until satisfactory evaluation results have been obtained.[1] The reflective action science methodology, followed by the research, is illustrated in Figure 1.

4. Diagnosis

We define the problem in our diagnosis as the possible attack strategies for an irrational attacker. To identify new potential attack strategies, we rely on a systematic review of Bitcoins design choices on the previously identified architectural layers (Section 2). For each layer, we define the goal and assumptions of our irrational attacker and review the layer's design choices. The design choices analyzed are only a subset of all properties associated with the layer; we restrict our review to only the components that have already been identified by the previous research as security

[1] As stated earlier, this is beyond the scope of this chapter, we envision that future work would assess the feasibilities of these attacks.

critical. This study's design choice restricts the complexity of reviewing a broad set of the Bitcoin design component. The rest of this section reviews each of the layers and state the goal and assumptions of the irrational attacker.

4.1 Protocol Layer

The protocol layer is considered the core of the Bitcoin network as the rules defined in the protocol drive the Bitcoin system. During the inception of Bitcoin, the protocol design was largely driven by the community of a few developers, and the choices made during the early days of Bitcoin persist in most cases. These decisions may not have considered all the attack vectors. Previous research studies have reported several issues related to the protocol layer design choices [53], [51], [50], [52]. Our study aims to identify the protocol layer decisions that may prompt security threats. We define the **Goal** and **Assumption** of irrational miner on protocol layer as follows:

Goal: *Identification and exploitation of design choices that may be prone to alteration to cause harm to the network.*

Assumption: *We assume that our irrational attacker is unable to change the protocol implementation for the participating nodes.*

4.2 Consensus Layer

The novel idea of bitcoin is the clever use of incentive based consensus algorithm [37]. As demonstrated in Section 2, most of the attackers theorized are consensus mechanism oriented such as the double spending attack. The security of the bitcoin network inherently depends on the quality of the consensus mechanism. This dependency may prove it crucial for an irrational attacker to attempt and exploit the consensus layer constructs. The attacker has the following goal and works under the following assumption:

Goal: *Identification and exploitation of design choices that may be prone to alteration by external factors such as computational power of the adversary.*

Assumption : *We also assume that the attacker has less than 50 % of the computational power of the network i.e.,* $\alpha < 50\%$. *We also reduce the complexity of analysis by not considering the network latencies involved in propagating information from the attacker to the rest of the network.*

4.3 Network Layer

Any computing system that is connected to a network may possess one of many identified network vulnerabilities [10]. As the bitcoin system is a peer to-peer system, most of the possible actions in the system execute over a networking environment. Due to the open nature of the bitcoin system, the networking component of the system is exposed to the public domain where any computing node can become a part of the network of bitcoin. This dependence and open nature make bitcoin a very lucrative target for networking attacks such as DOS (Denial of Service) [23]. We assume that the irrational attacker functions under the following goal and assumption:

Goal: *Identification and exploitation of networking attacks to reduce the reliability of the network by inducing latency.*

Assumptions : *We assume that our irrational attacker can directly establish a connection to a particular portion of the network.*

4.4 Economic Layer

The consensus mechanism of bitcoin assumes that most participants are rational and will be honest to network as the reward they receive for honest behavior is more significant than dishonest behavior. The reward in bitcoin is disseminated to the participant in the form of BTC. The intrinsic value of BTC has been argued upon in the research [30], [17], [16]. This argument further strengthens the dependence of Bitcoin on the economics associated with participation [36]. Prior research in the economics of Bitcoin has reported that it is indeed possible to manipulate the intrinsic value of the BTC [43]. The illustrated irrational attacker aims to exploit the economic vulnerabilities of the Bitcoin system with the following goal and assumptions:

Goal: *Identification of economic vulnerabilities that may be exploited given an irrational motive.*

Assumptions : *We assume that our irrational attacker has an unconstrained supply of monetary assets that the attacker may use to manipulate the network.*

Based on the goals and assumptions listed above using self-interpretation of the system layers result in the attacks identified in Table 1. Table 1 contains a non-exhaustive list of possible attacks by the irrational attacker. As demonstrated in numerous action research studies, the diagnosis of these attack vectors is mostly driven by self-interpretation of the problem [7], [4], [8]. The validity of these attacks is further evaluated in the subsequent phases of the action research plan (Section 3). The remainder of this section outlines the identified attacks in detail.

Table 1. Irrational attacks.

Layer	**Identified Attacks**
Protocol Layer	Nonce Attack and Difficulty Re-calibration Attack
Consensus Layer	Extortion Attack
Network Layer	Denial of Service Attack
Economic Layer	Panic Selling and Griefer Attack

4.5 *Protocol Layer Attacks*

4.5.1 Nonce Attack

Every block in the bitcoin's blockchain contains a 32-bit field called **Nonce**.

The value of this field is varied by the miners to find the hash of the block that fulfils the target requirement of the network. This calculation is performed by specialized mining equipment. The successful calculation may result in a reward. Even though the irrational attackers' aim is not to attain higher profit, but if the attacker can gain a profit greater than the hashing power proportion, it may impact the profit of others. We observe that the honest portion of the network is probabilistic able to mine the following BTC as a reward:

$$Reward_{Honest} = B_r * (100 - \alpha)/100 \quad (1)$$

If the attacker can successfully mine more blocks without increasing the α, it may reduce the value of $Reward_{Honest}$. The loss in the reward due to the attacker's ability to mine more blocks without a significant increase in α causes harm to honest participants. The proposed nonce attack is an attempt to exploit this in order to cause harm to the network.

We speculate that due to the endianness of the mining hardware, one type of hardware may favor finding odd values of nonce than even. If the attacker can establish a pattern of nonce value, it may significantly impact the attackers' possibility of finding the actual value of nonce, thus attaining a higher reward leading to the loss of others. **Attack Strategy:**

- *Step 1*: Retrieve historical nonce data from all blocks on the blockchain.
- *Step 2*: Observe the odd and even ratio of nonce over time to devise a new mining strategy to maximize the chances of mining the next block. The attacker may also examine other patterns with a nonce to observe any other numerical biases induced by the hardware used.
- *Step 3*: Test the new mining strategy in a private bitcoin deployment to observe any improvements.
- *Step 4*: Based on the results from Step 3, the attacker may alter the strategy devised in Step 2.

- *Step 5*: The new strategy is deployed on the actual network in an attempt to increase α, consequently resulting in a lower $Revenue_{Honest}$.

4.5.2 Difficulty Recalibration Attack

The Bitcoin network aims to, on average, generate one block every 10 minutes. To maintain this block creation time, the network periodically makes it harder or easier for the rest of the network to find a solution for the nontrivial mathematical problem. This change in the difficulty of mining is referred to as the difficulty of recalibration. In Bitcoin, the difficulty recalibration is performed after 2015 blocks [54].

The block creation time is often referred to as a bottleneck for the performance of the blockchain, which is often measured in transactions per second (TPS) [44], [31]. In contrast to the traditional payment systems, Bitcoin exhibits a low TPS speed. The low TPS has been speculated to be a barrier in the mainstream adoption of bitcoin as a payment system [40].

The irrational attacker can further slow the adoption by artificially manipulating the TPS of Bitcoin. We speculate that the time difference between two difficult recalibrations may be exploited to harm the network. We devise a novel strategy in which the attacker will first attain a significant α followed by sudden withdrawal from the network. We have illustrated the attack flow in Figure 2.

The attack is significantly limited by the security provision in the Bitcoin core client, which limits the maximum rise or fall in the value of the difficulty after every recalibration [22]. However, this limit does not omit the possibility of an attack over a more extended period. The abrupt change in the total hash power induced by the irrational attacker will cause the $Revenue_{honest}$ to fluctuate significantly over time, making it less reliable to mine. When the difficulty decreases after the attacker leaves the network, the block creation time will increase significantly due to the high difficulty. This increase in block creation time will subsequently

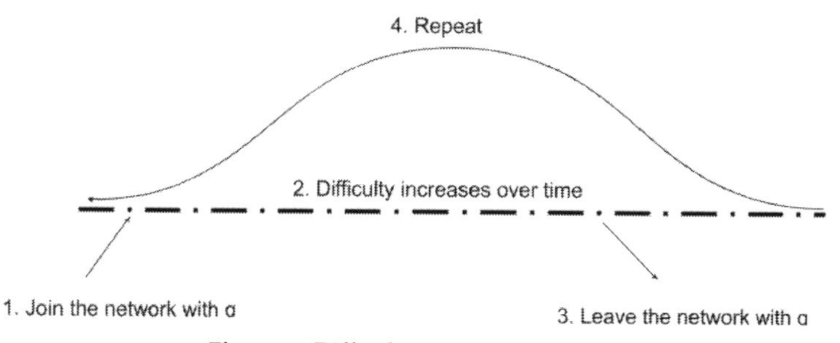

Figure 2. Difficulty recalibration attack.

result in a low TSP. As critiqued above, low TSP is highly undesirable as it causes direct harm to the adoption of Bitcoin as a payment system. **Attack Strategy:**

- *Step 1*: The attacker with hash power α joins the network with hash power (H_{total}) resulting in a new increased hash power ($H_{network} = H_t otal + \alpha$).
- *Step 2*: Assuming that the value of α is a significant portion of $H_{network}$, the sudden growth with result in a significantly low B_t due to a D that can only be changed after 2016 blocks. The attacker will wait till the difficulty is raised by a factor proportional to α.
- *Step 3*: Once the value of D has gradually increased to maintain a B_t of 10 minutes, the irrational attacker will decide to withdraw from the network.
- *Step 4*: The withdrawal of */alpha* from the network will result in a significantly lower $H_{network}$ but with the same D for the first recalibration period. This will result in a high B_t, thus a low TPS.
- *Step 5*: Once the network re-calibrates the difficulty to attain a B_t of 10 minutes, the attacker will start the process of B_t manipulation by repeating the attack from Step 1. The practicality of such an attack is analyzed in subsequent phases of the action research plan.

Theoretically, the attack can be carried out for an unbounded period leading to a very unreliable payment system with high variance in block creation. This may significantly harm network adoption.

4.6 Consensus Layer Attack

4.6.1 Extortion Attack

Gaining the majority in the Bitcoin network is considered hard due to the increasing cost of attaining mining hardware. One way of attaining a higher reward than the attackers hashing power proportion is Selfish Mining [32]. As reported in Section 2, in selfish mining, the attacker aims to get other miners to join his coalition to lose less reward. The rationale behind other miners joining the attackers' collocation is reduced loss in rewards. We propose another technique that aims to increase the profit of others while slowing down the network growth rate at the same time.

It has been reported that the transaction fee (T_f) only constitutes a tiny portion of total block reward [21], [19]. We propose exploiting this high dependence of consensus mechanism on the Block Reward (B_r) as the prime mean of reward dissemination. In line with previous attacks that have proposed using a high-value transaction to induce more forks in the network [38], we propose imposing a lower bound value on T_f for each block. The attacker may attempt to censor the network by only accepting transactions with $T_f > 1BTC$. This condition will result in only a tiny portion of the transactions being included in the block mined by the α proportion of the networks hashing power. Based on the value of α, the transactions backlogged may be significant, resulting in a low TPS rate without manipulating B_t.

We also speculate that other miners would be incentivized to join the attacker coalition as the attacker on average will earn more BTC from each block than honest miners as the lower bound value of T_f of the censored blocks will likely be higher than the honest blocks. The more significant the proportion of the network that participates in the attack, the more reward the participants will earn. This lucrative opportunity to earn more reward may incentivize the honest miners to join the attacker. **Attack Strategy:**

- *Step 1*: The attacker with hash power α will publically announce that the attacker will only include transactions that have a T_f higher than the lower bound defined by the attacker. This lower bound is higher than the moving average total T_f of past 100 blocks.
- *Step 2*: The attackers initializes the attack with its mining power.
- *Step 3*: The attacker will wait for others to join his attack. The public announcement may prompt rational miners to participate in the attack to earn more profit. As more rational honest miners join the coalition, the attackers α increases.
- *Step 4*: The attacker continues the attacks indefinitely causing the network to suffer a significant backlog of transactions slowing down the TPS, causing the Bitcoin network significant harm.

4.7 Network Layer Attack

Information system security is often evaluated based on the CIA triad model (Confidentiality, Integrity, and Availability) [11]. The desired security properties include confidentiality, i.e., the data processed or transmitted from the information system should not be exposed to unauthorized actors. The confidentiality in Bitcoin in ensured by the use of known secure cryptographic algorithms such as Elliptic-curve cryptography [2]. Another curial security property is the integrity of the data, i.e., the data transmitted or stored in the information system should not be altered and reminded in the intended form. This property is especially important in reference to Bitcoin as it aims to provide an immutable ledger of transactions. To attain this property, Bitcoin relies intensely on secure hashing function such as SHA256.

As Bitcoin is primarily a peer-to-peer networking system, the last of CIA triad property plays vital importance. Availability in a security context refers to the ability of the users to access the information and services provided by the information system. This property can easily be void in networking systems by inducing significant traffic to the networking device. The attacks on this property of the information system are known as Denial of Service attack, where the attacker aims to obstruct the service for a period of time [6].

As reported by [23], Bitcoin has implemented several safeguards to ensure that traditional DOS attacks do not result in loss of availability. This includes the blacklisting of bad actors in the system that sends out garbage block data [56]. We propose a new technique that aims to circumvent this limitation and exploit the node discovery protocol to induce noisy data in the network with the aim of slowing down the network. The node discovery protocol allows new nodes in the network to connect with existing nodes [27].

We aim to circumvent two security provisions against DOS implemented in Bitcoin Core Client [45]. The first provision aims to ensure that the connection to the DOS node is disconnected based on DoS score calculated by the client application for every connected node. The second provision bans the misbehaving nodes for 24 hours to ensure that the DoS attack is not successful. In our attack, we propose deploying lightweight bitcoin clients with just networking capabilities in a virtual environment. These virtual nodes actively connect with as many nodes in the network as possible and send garbage data to clog the networking link to the node. Once the virtual node is blocked by one of the peers, a new virtual node is deployed to connect to the peer with a different IP address and node address.

By virtualizing the process of node deployment, we can increase the number of full nodes on the network and attempt to connect to as many peers as possible on the network. These nodes can be deployed similarly to that of the one proposed in [45]. **Attack Strategy 1:**

- *Step 1*: If the total discoverable nodes present in the bitcoin network is *N*, the attacker will deploy greater than *N* virtual full nodes. Each virtual node will contain a list of all other virtual nodes to avoid a high degree of intercommunication.
- *Step 2*: After the deployment, the attacker's node will act as an honest node which only relays the messages between honest nodes while still connecting to as many nodes as possible.
- *Step 3*: After a predefined period, the attacker will start sending garbage data periodically to the connected peers.
- *Step 4*: As honest peers start blocking or disconnecting from the virtual nodes, the attacker will generate more virtual nodes that will attempt to repeat the process.

We speculate that such an attack may cause significant harm for its first iteration but will most likely be patched and does not constitute a longitudinal attack. To cause more harm to the network by DoS, the attacker may wish to exploit the vulnerabilities in Mining pools. The following text explores an alternate DoS strategy in which the target is Mining pools rather than the Bitcoin network itself.

As reported in [20], even deliberate DoS attack on the mining pool may yield a profit. A well-orchestrated strategy of DoS against mining pools by other mining pools may be profitable. They report that smaller mining pools have a greater possibility of earning a profit if they attack larger pools. We propose a variant of the deliberate attack to reflect the irrationality of the attacker. The irrational attacker is not concerned about the profitability and aims to cause harm to the network. We speculate that the attacker may reduce the total hashing power of the network significantly by reducing the effectiveness of several large mining pools. We propose the following attack strategy against mining pools:

- *Step 1*: The attacker may start by shortlisting target mining pools by identifying associated IP addresses or node addresses to communicate with. This can be done by maliciously joining the mining pool and connecting to as many honest miners in the mining pool as possible.
- *Step 2*: In this step, the attacker will attempt to congest the mining pool network by disseminating a large quantity of garbage data. The attacker can repeat this process until the mining pool blocks attackers node. In the case of being blocked, the attacker may similarly morph a new node to that of Attack Strategy 1.

4.8 Economic Attacks

4.8.1 Panic Selling

As reported in [31], the security of Bitcoin is largely dependent on the reward for mining. We observe that the real world value of the reward is a significant factor that provides intrinsic value to Bitcoin [39]. This dependence on the exchange rate of Bitcoin to fiat currencies renders it vulnerable to external manipulation. The price manipulation of Bitcoin has been studied in [42]. We now attempt to analyze it by assuming an irrational actor with the intent of harming the system. By reducing the exchange value of Bitcoin significantly, the attacker may compromise the honest majority because of a lack of incentive. We propose Panic Selling as a mechanism to harm the network, given that the attacker has substantial monetary assets. The proposed Panic Selling attack consists of the following steps:

- *Step 1*: The irrational attacker gradually accumulates a large amount of Bitcoin from numerous exchanges inducing an artificial need of Bitcoins. As bitcoin's market price is driven by supply and demand model [29], the attacker may increase the price significantly during the acquisition period.

- *Step 2*: Once the attacker has accumulated a substantial proportion of the Bitcoins in supply, a selling operation is performed. By suddenly dumping a large sum of Bitcoins, the attacker can reduce the exchange rate significantly. Depending on the acquisition power of the irrational attacker, the dumping may cause bitcoin to lack any incentive for participation.
- *Step 3*: The sudden drop in the price of Bitcoin may induce fear in stakeholder leading a panic selling, the attack may benefit from such a voluntary selling increasing the effectiveness of the attack.

Panic selling attack assumes that the attacker will sell the assets at the present market price and aims to drop the exchange rate by excessive selling. We propose another economic attack, in which the attacker sells Bitcoin at an irregular price to annoy other sellers and eventually reduce the value of Bitcoin.

4.8.2 Griefer Attack

In ludology, a griefer is defined as an action that acts irrationally to antagonize other participants [13]. We propose that the irrational attacker follow a Griefer pattern when conducting an economical attack to induce uncertainty in the environment. Uncertainty and its relation with the price have been widely studied [5] suggesting a strong reliance on price on the certainty of return. The exchange rate of Bitcoin to more traditional fiat currencies is largely decided by the exchange platform. These exchange platforms often tend to have some variance in the exchange rate of Bitcoin. We propose to exploit this delegation of exchange rate determination to the third party. In the speculated attack, the attacker will arbitrarily choose the price of Bitcoin to confuse the market of resell value. The attacker may follow the following attack strategy:

- *Step 1*: Similar to Panic Selling, the attacker must gather a significant amount of Bitcoin before attacking the network.
- *Step 2*: Based on the accumulated amount of Bitcoin, the attacker may wish to either sell the Bitcoin at an arbitrary price to buyers on buying-selling platforms or establish a new exchange that offers Bitcoins at a significantly low price.
- *Step 3*: The price difference may prompt users to buy Bitcoin from the newly established exchange rather than more traditional exchanges. This migration of potential buyers may lead to most exchanges adopting the low exchange rate dictated by the malicious attackers' exchange.

5. Conclusion

The security provisions of Bitcoin have been subjected to academic scrutiny since its conception in 2009. Numerous research articles have investigated the security characteristics of decentralized cryptocurrency. Due to the economics involved in the cryptocurrencies, numerous of these articles assume a rationale attacker with the scheme to maximize the profit. This conjecture omits the possibility of an irrational attacker with the purpose of harming the cryptocurrency irrespective of the cost.

This chapter presents various novel attack vectors that an irrational attacker may exploit to jeopardize the bitcoin network. The paper manifests attacks on four facets: Consensus layer attacks, Network layer attacks, Protocol layer attacks, and Economic attacks. By modelling an irrational attacker, we can investigate the plausibility of a potential large-scale organizational or governmental attack on the Bitcoin network.

We have identified nonce and difficulty recalibration attacks as two potential sources of irrational attacks on the protocol layer. Both of these strategies have not yet been seen as an attack vector. Both of the attacks seem theoretically viable however it would be worthwhile to explore these attack vectors as these can likely be fixed by a protocol level fix.

Another type of attack that we have identified relies on the consensus protocol of the network. Other attacks that can induce instability in the network include network and economic attacks. For instance, panic selling can induce a drop in the exchange rate of Bitcoin that could potentially impact the profitability of operating in the network.

5.1 Future Work

The purpose of this chapter is to illustrate the possibility of irrational attacks on Bitcoin and other cryptocurrencies. We believe that these strategies need to be examined further to rule out any potential threat these attack strategies may pose to the network.

We intend on constructing an experimental setup to test the feasibility of these attack vectors. Specifically, examining the possibility of economic attacks using reinforcement learning may allow us to understand the economic dynamics present in the crypto-economies.

Glossary

Bribery attack is an attack in which the attacker intends to double spend a transaction t_d by incentivizing other miners to favor the fork that includes the conflicting transaction.. 7

Eclipse mining is an attack in which the attacker partitions the victim from the rest of the network.. 7

PoW Proof of work (PoW) is a method of securing a cryptocurrency network and confirming transactions by having computer networks verify large amounts of data.. 4

Selfish mining is an attack in which the attacker aims to collect an unfair share of reward by causing harm to the other participants of the network.. 6

Whale transaction is defined as a transaction with very high monetary value as the transaction fee, which may promote forks in the network.. 7

References

[1] Roger, D. Evered. 1978. An assessment of the scientific merits of action research Gerald 1. Susman and. In: *Administrative Science Quarterly*, 23.4: 582–603.

[2] Neal Koblitz. 1987. Elliptic curve cryptosystems. In: *Mathematics of Computation*, 48.177: 203–209.

[3] Harold, G. Levine and Don Rossmoore. 1993. Diagnosing the human threats to information technology implementation: A missing factor in systems analysis illustrated in a case study. In: *Journal of Management Information Systems*, 10.2: 55–73.

[4] Richard, L. Baskerville and Trevor A. Wood-Harper. 1996. A critical perspective on action research as a method for information systems research. In: *Journal of Information Technology*, 11.3: 235–246.

[5] Raymond Deneckere, Howard P. Marvel and James Peck. 1997. Demand uncertainty and price maintenance: Markdowns as destructive competition. In: *The American Economic Review*, pp. 619–641.

[6] Christoph, L. Schuba et al. 1997. Analysis of a denial of service attack on TCP. In: *Proceedings. 1997 IEEE Symposium on Security and Privacy (Cat. No. 97CB36097)*. IEEE., pp. 208–223.

[7] Richard, L. Baskerville. 1999. Investigating information systems with action research. In: *Communications of the Association for Information Systems* 2.1: 19.

[8] Richard, Baskerville and Michael D. Myers. 2004. Special issue on action research in information systems: Making IS research relevant to practice: Foreword. In: *MIS Quarterly*, pp. 329–335.

[9] Rikard Lindgren, Ola Henfridsson and Ulrike Schultze. 2004. Design principles for competence management systems: a synthesis of an action research study. In: *MIS Quarterly*, pp. 435–472.

[10] Simon Hansman and Ray Hunt. 2005. A taxonomy of network and computer attacks. In: *Computers & Security* 24.1: 31–43.

[11] Sattarova, Y. Feruza and Tao-hoon Kim. 2007. IT security review: Privacy, protection, access control, assurance and system security. In: *International Journal of Multimedia and Ubiquitous Engineering*, 2.2: 17–32.

[12] Satoshi Nakamoto et al. 2008. Bitcoin: A peer-to-peer electronic cash system. In.

[13] Don Gotterbarn and James Moor. 2009. Virtual decisions: Video game ethics, Just Consequentialism, and ethics on the fly. In: *ACM SIGCAS Computers and Society* 39.3: 27–42.

[14] Ghassan Karame, Elli Androulaki and Srdjan Capkun. 2012. Two bitcoins at the price of one? Double-spending attacks on fast payments in bitcoin. *In: IACR Cryptology ePrint Archive*, 2012.248.

[15] Ghassan, O. Karame, Elli Androulaki and Srdjan Capkun. 2012. Doublespending fast payments in bitcoin. pp. 906–917. In: *Proceedings of the 2012 ACM conference on Computer and communications security*. ACM.

[16] Fran Velde et al. 2013. Bitcoin: A primer. In.

[17] Adrian Blundell-Wignall. 2014. The Bitcoin Question. In.

[18] Nicolas, T. Courtois and Lear Bahack. 2014. On subversive miner strategies and block withholding attack in bitcoin digital currency. In: *arXiv preprint arXiv:1402.1718*.

[19] Nicolas Houy. 2014. The economics of Bitcoin transaction fees. In: *GATE WP* 1407.

[20] Benjamin Johnson et al. 2014. Game-theoretic analysis of DDoS attacks against Bitcoin mining pools. In: *International Conference on Financial Cryptography and Data Security*. Springer, pp. 72–86.

[21] Kerem Kaskaloglu. 2014. Near zero Bitcoin transaction fees cannot last forever. In.

[22] Karl, J. O'Dwyer and David Malone. 2014. Bitcoin mining and its energy footprint. In.

[23] Marie Vasek, Micah Thornton and Tyler Moore. 2014. Empirical analysis of denial-of-service attacks in the Bitcoin ecosystem. pp. 57–71. In: *International conference on financial cryptography and data security*. Springer.

[24] Beat Weber. 2014. Bitcoin and the legitimacy crisis of money. *In*: *Cambridge Journal of Economics* 40.1: 17–41.

[25] Joseph Bonneau et al. 2015. Sok: Research perspectives and challenges for bitcoin and cryptocurrencies. pp. 104–121. In: *2015 IEEE Symposium on Security and Privacy*. IEEE.

[26] Ethan Heilman et al. 2015. Eclipse attacks on bitcoins peer-to-peer network. In: *24th {USENIX} Security Symposium ({USENIX} Security 15)*, pp. 129–144.

[27] Andrew Miller et al. 2015. Discovering bitcoins public topology and influential nodes. In: *et al.*

[28] Joseph Bonneau et al. 2016. Why buy when you can rent? bribery attacks on bitcoin consensus. In.

[29] Pavel Ciaian, Miroslava Rajcaniova and dArtis Kancs. 2016. The economics of BitCoin price formation. In: *Applied Economics* 48.19: 1799–1815.

[30] Anne Haubo Dyhrberg. 2016. Bitcoin, gold and the dollar–A GARCH volatility analysis. In: *Finance Research Letters* 16: 85–92.

[31] Arthur Gervais et al. 2016. On the security and performance of proof of work blockchains". In: *Proceedings of the 2016 ACM SIGSAC conference on computer and communications security*. ACM, pp. 3–16.

[32] Ayelet Sapirshtein, Yonatan Sompolinsky and Aviv Zohar. 2016. Optimal selfish mining strategies in bitcoin. In: *International Conference on Financial Cryptography and Data Security*. Springer, pp. 515– 532.

[33] Mark Walport. 2016. *Distributed ledger technology: beyond blockchain. UK Government Office for Science*. Tech. rep. Tech. Rep.

[34] Karl Wu¨st and Arthur Gervais. 2016. *Ethereum eclipse attacks*. Tech. rep. ETH Zurich.

[35] Samiran Bag, Sushmita Ruj and Kouichi Sakurai. 2017. Bitcoin block withholding attack: Analysis and mitigation. In: *IEEE Transactions on Information Forensics and Security* 12.8: 1967–1978.

[36] Jonathan Chiu and Thorsten V. Koeppl. 2017. The economics of cryptocurrencies–bitcoin and beyond. In: *Available at SSRN 3048124*.

[37] Nicola Dimitri. 2017. Bitcoin mining as a contest. In: *Ledger* 2: 31–37.

[38] Kevin Liao and Jonathan Katz. 2017. Incentivizing blockchain forks via whale transactions. In: *International Conference on Financial Cryptography and Data Security*. Springer, pp. 264–279.

[39] Dirk, G. Baur, Kihoon Hong and Adrian D. Lee. 2018. Bitcoin: Medium of exchange or speculative assets? In: *Journal of International Financial Markets, Institutions and Money*, 54: 177–189.

[40] Mauro Conti et al. 2018. A survey on security and privacy issues of bitcoin. In: *IEEE Communications Surveys & Tutorials*, 20.4: 3416– 3452.
[41] Ittay Eyal and Emin Gu¨n Sirer. 2018. Majority is not enough: Bitcoin mining is vulnerable. In: *Communications of the ACM*, 61.7: 95–102.
[42] Neil Gandal et al. 2018. Price manipulation in the Bitcoin ecosystem. In: *Journal of Monetary Economics*, 95: 86–96.
[43] John, M. Griffin and Amin Shams. 2018. Is bitcoin really un-tethered? In.
[44] Ashish Rajendra Sai, Jim Buckley and Andrew Le Gear. 2018. Optimal block time for proof of work blockchains. In.
[45] Bitcoin. *Bitcoin Core Implementation Github*. 2019. url: https:// github.com/bitcoin/ bitcoin.
[46] Ashish Rajendra Sai, Jim Buckley and Andrew Le Gear. 2019. Assessing the security implication of bitcoin exchange rates. In: *Computers & Security*, 86: 206–222.
[47] Ashish Rajendra Sai, Andrew Le Gear and Jim Buckley. 2019. Centralization threat metric. In.
[48] Ashish Rajendra Sai. 2021. Towards a holistic assessment of centralization in distributed ledgers. In.
[49] Ashish Rajendra Sai et al. 2021. Taxonomy of centralization in public blockchain systems: A systematic literature review. In: *Information Processing & Management*, 58.4: 102584.
[50] Gavin Andresen. *Block v2, Height in Coinbase*. url: https://github.com/bitcoin/bips/ blob/master/bip-0034.mediawiki.
[51] Gavin Andresen. *March 2013 Chain Fork Post-Mortem*. url: https: //github.com/bitcoin/ bips/blob/master/bip-0050.mediawiki.
[52] *CVE-2013-2273*. url: https://nvd.nist.gov/vuln/detail/CVE2013-2273.
[53] *CVE-2013-5700*. url: https://nvd.nist.gov/vuln/detail/CVE2013-5700.
[54] *Difficulty*. url: https://en.bitcoin.it/wiki/Difficulty.
[55] Finney, H. *The Finney attack(the Bitcoin Talk forum)*.
[56] *Weaknesses*. url: https://en.bitcoin.it/wiki/Weaknesses\ #Denial_of_ Service_.28DoS.29_attacks.

3

A Blockchain-Based RegTech System for Product Safety Enforcement

A Case Study of Food Import in China

Christophe Viguerie[1,*] and *Robert M. Davison*[2]

1. Introduction

Food safety is a major area of concern for all countries and its regulation is beneficial to international trade [37]. With the globalization of trade, food supply chains have become more diverse and complex, necessitating tighter food safety regulatory frameworks [5]. Countries' regulators are often brought into action when food safety incidents occur which then leads to stricter regulations [8,45]. The Chinese government has gradually strengthened its legal provision with respect to food import over the past seventy years (see Appendix A). In recent years, the implementation of a more stringent regulatory framework was influenced by both domestic incidents [35], and the internationalisation of regulation [9]. Moving from a food hygiene to food safety regulation environment, the central government in China has developed stricter measures over the past two decades for both national production and imported products. This led to

[1] DBA Programme, City University of Hong Kong, Tat Chee Avenue, Kowloon, Hong Kong.
[2] Dept of Information Systems, City University of Hong Kong, Tat Chee Avenue, Kowloon, Hong Kong.
Email: isrobert@cityu.edu.hk
* Corresponding author: cv.cityu@gmail.com

a more complex legal environment for businesses interested in exporting food to China. Despite a low level of refusal at the border, importing food into China continues to be a complex process [15].

Following the 2008 global financial crisis, strict regulations put tremendous pressure on financial institutions that led to the use of regulatory technology (RegTech) to ease the compliance process [1]. With the more recent introduction of regulatory policies related to customer due diligence, requiring companies to "know their customers" as part of the anti-money laundering environment, blockchain technology appeared as a RegTech with the capability to affect potential improvements in the process and ease the compliance burden [30]. Beyond this case, blockchain technology is increasingly considered as a solution to support regulatory compliance [17]. However, research to date on regulations largely focuses on analysing government decisions rather than enforcement [16]. In this chapter, we study the potential of a blockchain-based RegTech system for food safety enforcement and import in China.

The case studied is the business-to-business (B2B) food platform 'Foodgates' that markets French food products to Chinese buyers. Foodgates opens the opportunity for small producers in France to export their products to Chinese consumers by connecting them to potential buyers and managing the whole process of transportation door-to-door. Their motto, "Because it's complicated to export to China", highlights the acute challenges experienced by small businesses in complying with China's food import regulations. Benefiting from its in-between position, Foodgates collects all relevant product information that is required to satisfy China's regulations. Additional information is collected during the transportation process that is also managed by Foodgates. Information is safely stored on a blockchain which facilitates the compliance process. However, the substantial paperwork load required by the customs and other regulatory administrations in China still prevents the optimal use of blockchain as a regulatory technology. Despite this situation, blockchain has been embraced by the central authorities in China as a strategic technology to develop e-government nationwide. This decision paves the way for wider adoption of blockchain to deal with all government services including customs and import procedures. In this chapter, we provide a detailed case of how blockchain can support product safety enforcement.

2. Food Safety Regulation in China

2.1 Internal and External Factors

Since 1949, the food regulation system in China has been continuously adjusted to the circumstances and challenges (for more details on the regulatory evolution see [16],[18],[29]). In recent years, food safety regulation in China has been driven by both the global regulation convergence and internal food safety problems. The globalisation of the economy has resulted in a global regulation movement pushing trade partners to harmonise their regulations. That convergence has greatly influenced China's regulatory measures especially for the deemed non-strategic industries [9]. In particular, international food safety standards have greatly influenced China's policies over the past twenty years. For instance, a key motivation was the repeated occurrence of safety issues related to China's food exports in the late 1990s that were putting the country's role as a food exporter in jeopardy [7].

China has also witnessed severe food scandals over the past two decades that have made necessary a strengthened regulatory regime for domestic production [22,35]. It is worth noticing that the food regulation has been enforced with a higher degree of compliance for exported products than for local production consumed within the country [7,9]. The full compliance for export food products meets international standards such as the ISO and Codex. This more stringent enforcement policy applies also to imported food and, in that respect, the General Administration of Customs (GACC) has strengthened its inspection procedures at all ports of entry in the country following 2018 regulations [9].

2.2 From Food Hygiene to Food Safety Regulation

The implementation in June 2009 of the first Food Safety Law translates China's decision to recognise food safety as a public health priority [29]. As explained by Liu, Mutukumira and Chen [29], the establishment of the Food Safety Committee the following year confirms the government's priority change from food hygiene to food safety (see Appendix B). The objective is to create a unified and centralised governance mechanism able to ensure food safety from "farm to table" which was reinforced further by the 2018 reform that instituted the State Administration as the top body for food safety enforcement and regulation [18]. Even if food safety and its regulation are a global challenge, the situation in some developed countries like United States, members of the European Union and Japan seem to be more adequate than in China [18]. Nevertheless, the Chinese government has reformed its system several times and improved drastically its effectiveness over the past decade.

The more stringent regulation toward imported food responds to the increasing inflow of foreign food products. This increase reflects the need of China for some products that are not currently produced in the country, or at least not in sufficient quantity or with adequate quality [24]. Besides, local food incidents have increased the attractiveness for imported goods as consumers want safer products [26]. In addition, the evolution of living standards enables a larger part of the population to purchase such imported food products that become part of higher-class consumption habits [15]. All of this contributes to make China one of the world's largest food importers [9]. Therefore, it is important to analyse how the food safety regulation impacts food imports in China.

2.3 Impact on Food Import in China

Over the last decade, the Chinese government has deployed some highly deterrent measures to reduce the risk of food incidents to consumers. This tighter regulation applies also to the import of foreign food products. In March 2021, the U.S. Department of Agriculture published a report on "China's Refusals of Food Imports" [15]. It appears that, even though the food imports' average rate of refusal by Chinese authorities was only 1 percent over the period from 2006 to 2019, compliance with the current regulatory system remains challenging. Although limited in percentage, the refusal decisions affect thousands of shipments.

The latest regulation reform focused on improving product traceability supported by labelling and country of origin certification requirements for imported food [26]. Documentation, record-keeping, and tracking are also among the requirements imposed on imported foods and represent a significant reason for lack of compliance or violation of the regulation that leads to import refusal [15]. Food exporters to China need support to ease the regulation compliance burden and applying blockchain could provide a solution.

3. Blockchain-Based Food Safety Compliance System

3.1 Blockchain Technology as a Regulatory Technology

Following Beck, Avital, Rossi and Thatcher [4], blockchain technology's key features are a cryptographically secured distributed ledger and a consensus mechanism that rules the transaction validation process. The ledger records chronologically the consensually validated transactions and is securely distributed to all blockchain participants [36]. As a ledger's data are accessible by all participants, the system is transparent [19]. In addition, data are protected by cryptography and thus are immutable which makes their records trustworthy [3]. In substance, the blockchain system delivers trustworthy and verifiable records directly to participants without the involvement of any third party trusted entities [11].

RegTech was originally designed to simplify and facilitate companies' compliance with regulations, in particular in the aftermath of the 2008 global financial crisis [1]. RegTech is defined by De Filippi and Hassan [12, p.15] as "technology that can be used both to define and incorporate legal or contractual provisions into code, and to enforce them irrespective of whether or not there subsists an underlying legal rule". More recently, RegTech has been integrated into a reconceptualised regulatory regime with the purpose of providing a higher level of service. This application is driven by governmental institutions such as the Bank of England that recently tested the use of new digital technologies (distributed ledger technology, artificial intelligence, and machine learning) to facilitate compliance by making regulations clearer and providing regulators with real time information [31]. To date, most RegTech applications of blockchain have been made in the financial sector [17,34]. In this chapter, we consider the potential application of a RegTech system based on blockchain technology that facilitates food safety enforcement.

3.2 Applications of Blockchain for Food Safety

Food safety is a public health concern and the prerogative of the governments that act as regulators. Rules and regulations must be complied with by all actors aiming at selling, importing, or exporting products in a particular territory. In this chapter, we look specifically at the food safety enforcement for foreign exporters aiming at exporting goods to China.

Blockchain is seen as a promising technology whose properties can bring transparency and enhance efficiency in a business environment [25]. In particular, blockchain technology creates a trusting environment among supply chain participants that facilitates information sharing, avoids reconciliations and therefore enhances process efficiency [27]. The trust in the ledger replaces the need for intermediaries. Blockchain works as a solution to enable transparent tracking of information throughout an economically, socially, and environmentally complex supply chain environment [38,44].

Based on these key properties, some researchers describe blockchain as a recordkeeping technology [28]. This aspect is instrumental in the use of blockchain for provenance. The data immutability not only provides a tamper-proof record of archived information, but also a trace of who validated the record and when it was done. This is particularly important for record authenticity, which is instrumental in regulation compliance. For example, information about provenance, certification of origin and quality labels are required by the Chinese authorities for imported food. By providing a transparent and immutable record, blockchain can facilitate inspections [14]. Blockchain can also increase trust in food safety by providing transparency and reducing the severity of security issues [37]. In summary, the technology can support food safety from "farm to table" [42]. Nevertheless, the actual use of blockchain for facilitating product safety enforcement also depends on the current blockchain environment in China.

3.3 Blockchain Environment in China

China has adopted a dual approach to blockchain. First, it restricted and banned activities deemed illegal.[1] Later, it acknowledged the benefits of this innovation in supporting a more transparent and efficient government [6]. In the latter sense, blockchain could replace government registries and enable a global trust in public records [28]. In 2016, China's central authorities promoted the adoption of standards to support development and application of blockchain in the economy [21].

Thus, in the 13th Five-Year Plan, the Chinese State Council launched the Trusted Blockchain Open Lab to embrace blockchain as a technology for improving people's lives but rejected cryptocurrencies [13]. This stance was confirmed in the 14th Five-Year Plan (2021–2025) that supports the development of blockchain for government nationwide [6]. The advanced involvement of Beijing Municipality shows already how blockchain can greatly enhance a large part of government services [41]. The recent development of national standards for blockchain application by the Chinese central authorities also paves the way for a China blockchain [6]. In addition, in 2020 the Chinese government officially launched the Blockchain-based Service Network (BSN). This blockchain is managed by the government and is designed to support the adoption of blockchain across the country [40].

4. Research Framework and Methodology

4.1 Conceptual Framework

Ogus [32] has stressed that organisations are willing to initiate expenditure to gather requested information with the objective of complying with the law or regulations. Failure to meet regulatory requirements in terms of information provision potentially puts organisations at risk. This risk is referred to as error costs and arises from unexpected legal decisions, legislation misunderstanding by organisations or lack of clarity of legal rules [32]. Compliance with food safety regulations is particularly complex, as different standards, related to the product itself and its production process, translate into regulatory codes and guidelines which require a wide range of information from the producers and possibly prior approval of products [33]. It is therefore essential for food exporters to ensure documents provided to customs and other regulators enable full compliance with foreign food safety regulation. In that respect, Henson and Heasman [20] propose a compliance process model that supports

[1] https://www.chinabankingnews.com/2017/09/04/china-declares-initial-coin-offerings-illegal/.

food businesses to deal with new food safety regulations and conclude that small firms are more at risk of failing to comply.

Several researchers highlighted the positive contribution of blockchain technology to supply chain management. Among others, blockchain properties enhance supply chains' transparency and facilitate the sharing of information [38,44]. Besides, some information required by food safety regulators necessitates a developed product traceability [26]. There, blockchain technology provides scope for a thorough and transparent information system from fields to kitchen [39]. Furthermore, some research articles acknowledge that blockchain reaches the status of RegTech with several cases in the financial field [1,12,17].

The objective of this research is to understand how blockchain, as a RegTech, can contribute towards supporting food exporters to comply with foreign food import regulations and facilitate food safety enforcement. Figure 1 depicts this first conceptual relationship, where arrows represent information flows.

In this chapter, we are looking at a particular case of this conceptual framework, a blockchain-based information system that can serve as a RegTech for French food producers that export their products to China. In addition to what was established in the literature review, Appendix C highlights the complexity of the food safety regulatory system in China for foreign exporters. Six China administrative agencies are involved in regulating and enforcing food safety regulation (for a complete list and scope of each agency, see https://www.transcustoms.com/guide/). Prior to the shipment, the foreign exporter needs to apply for GACC registration, and the documentation required to clear customs must be submitted. Document requirements can vary from one product to another, but the following documents must be prepared: commercial invoice, detailed packing list, bill of lading, certificate for export from country of origin, hygiene/health certificate, sample of original label, sample of Chinese label, and inspection certificate issued by the General Administration of Quality Supervision, Inspection & Quarantine (AQSIQ).[2] During the customs process, all these documents will be reviewed, and the Chinese

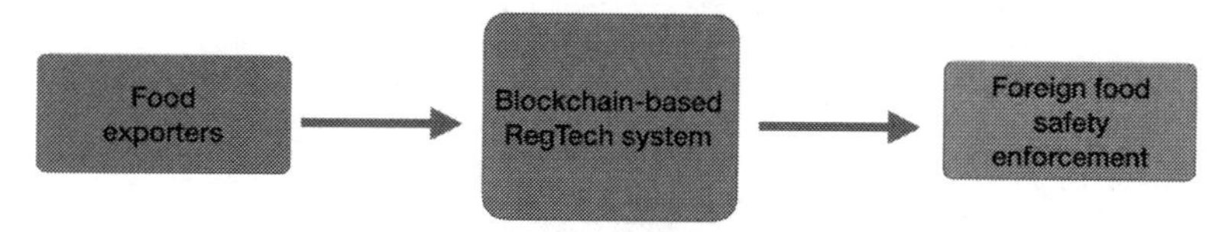

Figure 1. Conceptual framework of a blockchain-based RegTech system for food safety enforcement.

[2] https://www.china-briefing.com/news/exporting-food-products-to-china-regulation-and-procedure/.

label will be checked. Then, a food sanitary inspection can be conducted before clearance.

The complexity of the process and the regular modification of the regulations in China that has also been stressed in Section 2 create a challenging environment for food foreign producers, especially for small size organisations. In that respect, we study the case of a trade platform that applies blockchain technology to manage product and transportation information and support food exports from France to China. The objective is to answer the following question: How can a blockchain-based RegTech system facilitate both China's food import regulation enforcement and French producers' export process? We address this research question by understanding the structure of the system and by following a real case import process. The conceptual framework presented in Figure 2 shows information flows (numbers 1, 2 and 3) and the central role of the platform's blockchain-based information system as an intermediary between the French producers and the China food import authorities. The thin arrows represent the physical transportation of products from France to China.

There are two information sources. One from the French producers (flow 1) and another one during the transportation (flow 2). Data of flow 1 are related to product information provided by the French producers and are required by Chinese authorities for any food import. Data of flow 2 might have different origins as they are captured during the product transportation process and are related to logistics information (movements of the shipment) and conditions of the products during the transportation (temperature and others). The platform collects these pieces of information and stores them in its blockchain-based information system.

Then, the blockchain-based information system makes information available to China's food import authorities during the import process. We analyse the nature of information collected in flows 1 and 2 and how

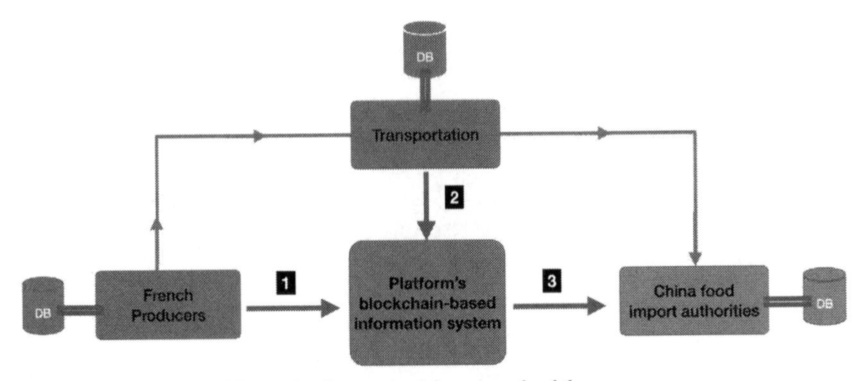

Figure 2. Conceptual framework of the case.

Foodgates' blockchain-based information system supports both the import process and China food safety enforcement in flow 3. We then conclude by discussing the potential of a blockchain-based RegTech system for facilitating China's food import regulation enforcement.

4.2 Research Methodology

The elements related to the methodology and the method developed in this chapter are inspired by the guidelines, designs and methods from Creswell [10], Baxter and Jack [2] and Yin [46]. We have also considered the specific recommendations of Treiblmaier [43] on blockchain case studies.

The conceptual framework that emerged from the literature review supports the positive contribution of blockchain as a regulatory technology in the finance field. The motivation of this research is to analyse the potential of blockchain in facilitating regulatory compliance in a product safety enforcement context. This is a contemporary phenomenon and the number of companies that have implemented such a system outside of the financial sector is still very limited. To understand this complex phenomenon, we must conduct an extensive and in-depth analysis of the phenomenon and this analysis necessitates a holistic and real world perspective. The case study, which is an empirical method that suits contemporary phenomenon in-depth study within a real world context, emerges as the most fitting method for this research. As an epistemological orientation, we consider there is a single reality, and this reality is independent of any observer. This realist perspective applies to the descriptive case study described in the following section.

To answer the research question, we specifically describe an information system based on blockchain technology that supports food import compliance. This single case design is holistic as there is a single unit of analysis. The unit of analysis is considered as a critical case that can provide data linked to the research question. The scope of the single case studied in this research is the platform developed by the company Foodgates.

Foodgates describes itself as "Foodgates is the only fully transparent food B2B platform with local presence in France and China, audited producers and products of excellence ready to be imported in China". Foodgates selects the producers and the products, works closely with certification bodies and Chinese authorities such as CIQ (China Inspection and Quarantine) to make sure all products on the platform comply with Chinese regulations. Foodgates provides all supply chains and transportation support from producers to table. Foodgates is a fully transparent platform with all pieces of information related to provenance, transportation and custom clearance being stored on a blockchain.

Prior to conducting the fieldwork, general information about Foodgates platform was captured from the website (https://foodgates.cn). Then data are collected from different sources such as companies' archives, observation of a transaction and semi-structured interviews. The persons who form the main group and represent the immediate topic of our case study are the company's employees both in France and in China. Outside of this group, companies that have developed the software solution and the blockchain system are part of the context of our case study and can support the understanding of the technological process. The French producers, the Chinese authorities related to food safety regulation and enforcement, and the Chinese consumers are not part of this case study's scope.

The case study leads to an explanation related to how to design an information system based on blockchain technology that facilitates compliance with food safety regulations in China. This descriptive analysis of a real world case provides a conceptual framework that enables the drawing of generalisable conclusions. The analysis of a real case import flow can validate or stress the limitations of such a blockchain application. By comparing the findings against the theoretical statement, the statement can possibly be revised. This revision can be motivated by issues intrinsic to the blockchain system itself or by external factors.

5. Case Study: Foodgates

5.1 Interviewees

The following persons have been interviewed as part of the study.

- Mr. Igal Chreky is Vice President of ASI Group and is based in Shanghai. Igal Chreky was one of Foodgates project's leaders. At the early stage, Igal Chreky defined blockchain system's needs and requirements, fixed medium and long-term traceability objectives of the platform, initiated the discussions with Younicorns and developed the relationship with VeChain and DNV.
- Ms. Hebe Liu is Foodgates Operation Manager in China and is based in Shanghai. Hebe Liu was previously working in the Food and Beverage (F&B) department of ASI Group in China. She was involved in Foodgates project and supported the process definition on the platform.
- Mr. Mathieu Borgé is Foodgates Sales and Marketing Director and is based in France. Mathieu Borgé was previously in charge of the French-related activities of ASI Group. He is co-founder of Foodgates and was part of the initial ASI Group project's team in charge of the French part of the activity.

- Mr. Jonathan Horyn is Younicorns co-founder and Foodgates Chief Technical Officer (CTO), based in France. Jonathan Horyn has been involved at the early stages of the Foodgates project and is responsible for the platform system and blockchain aspects such as traceability tracking, product detail displaying and interface with the trading platform.
- Mr. Jérôme Grillères is VeChain Tech General Manager Europe and is based in France. Jérôme Grillères was involved in the November 2019 operation in Shanghai and partnership with Foodgates and DNV.

In Appendix D, we provide an overview of the different companies involved in the Foodgates project.

5.2 Foodgates Project Origination

Original business problem

Originally, Carrefour and Auchan, two French multinational retail corporations, requested the support and experience of ASI Group to import their respective private labels in China. The private labels were made of multiple F&B products of different natures and produced at different places in France. ASI Group was able to provide four key services:

1. Regulatory compliance: as these private label products had never been imported to China, ASI Group performed importability tests and listed all regulatory elements that were needed to be added to comply with Chinese regulations (additional label, modification of the packaging, certifications, documentation, etc.).
2. Import licenses: ASI group used their licenses to import products for Carrefour and Auchan in China.
3. Transportation: Carrefour and Auchan needed a service provider for their private labels to coordinate the delivery from several warehouses in France and organise shipments to China. Transportation was mostly carriage by sea. The service also covered the customs clearance and delivery to Carrefour and Auchan's warehouses in China.
4. Payment in Chinese Yuan: ASI Group was also able to invoice the service in Chinese Yuan to the Chinese subsidiaries of Carrefour and Auchan.

The set-up of all these services was overseen by ASI Group F&B department and involved a team in China with Hebe Liu and another team in France led by Mathieu Borgé. In order to cope with the important flow of transactions and to support the import process, ASI Group started to use digitised information.

Development of Foodgates platform

Extending on this first project, ASI Group then decided to develop a global platform under the code name Food Trading Platform (FTP). Igal Chreky said, "The objective was to connect with as few intermediaries as possible; from European small to medium-sized enterprise (SME) food producers to Chinese professional buyers. The solution was to develop a fully integrated end-to-end solution serving as a business-to-business (B2B) platform from the farm in France to the distributor's warehouse in China". This platform should also be able to provide for digitisation of all pieces of information related to a transaction between a seller (food producer) located in a particular country and a buyer located in another country. The flow of information included information about the transaction (seller, product sold and buyer), information from the origin and destination customs, and information captured during the transportation process from the production place to the buyer's warehouse. The payment process was also included at the end of the process. Figure 3 describes the process steps and the different sources and flows of information.

FTP became Foodgates and was launched during the Shanghai China International Import Expo (CIIE) in November 2019. During this important trade fair, the Chinese and French presidents could enjoy a piece of "Label Rouge" Limousin beef imported through the Foodgates platform whose origin was certified by DNV and traceability available

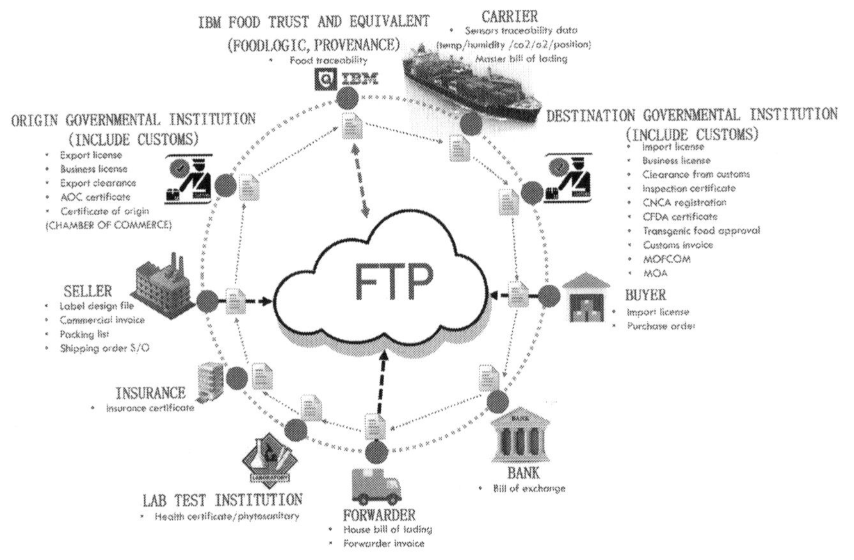

Figure 3. FTP project. Source: with permission from ASI Group.

on the VeChain blockchain (see Figure 4 here below). Foodgates platform was introduced as the first public and disruptive blockchain ecosystem in the food industry for French producers and business Chinese buyers who can access certified information throughout the whole value chain, from the production in France up to the delivery to their warehouse (Foodgates press release, 2019).

The event of November 2019 was a foundational event. At that time, the marketplace did not yet exist. It was an opportunity to partner with L'Association Limousine de la Qualité et de l'Origine,[3] that promotes beef of Limousin origin worldwide, and to present the traceability solution using blockchain technology. That was a milestone in the Foodgates project and confirmed that blockchain was to be a fundamental element of the platform. The name Foodgates was created for this specific event. The B2B platform was in development and this event confirmed the validity of the concept.

After incubation by ASI Group, the company Foodgates SAS was registered in France in March 2020. The Foodgates platform was then launched in November 2020. Currently, Foodgates operates from France and with a downstream operation team based in China that is focused on sales and logistics.

Figure 4. Picture of President Xi of China and President Macron of France taken at Shanghai China International Import Expo (CIIE) 2019.

[3] https://www.label-viande-limousine.fr/limousin-promotion.html.

5.3 Foodgates Platform Process Flow

Figure 5 illustrates the different steps pertaining to a transaction. In the first step, French producers are registered on the platform. Generic information about the company and the products are required which enables the creation of the profile. Once registered, the company then indicates a category and a description of products to be exported via the Foodgates platform. The main categories listed are dairy, meat, sea food, wine & spirits, grocery, beverage, and bread & pastry. Following this, a systematic importability check is made to ensure that the product can legally be imported into China. Mathieu Borgé noted, "We guarantee Chinese buyers that all products displayed on Foodgates platform can be imported to China. We are not just a marketplace but an added-value digital platform". Next, products are displayed on the platform with an indication of price and minimum order quantity. A Chinese client can then order a product by indicating a quantity and selecting the International Commercial Terms (Incoterm). After checking the product's availability with the producer, the Chinese buyer receives a confirmation of the price which includes transportation up to the warehouse and customs clearance. When the Chinese buyer accepts this confirmation, the transaction is validated on the platform and a QR code is created that enables to track all documents and information related to this transaction.

The transportation process follows. During this process, several types of information are collected on the platform: transportation documents, GPS tracking of the shipment that enables a timestamping of the different steps and a temperature check of the product (when applicable). Using

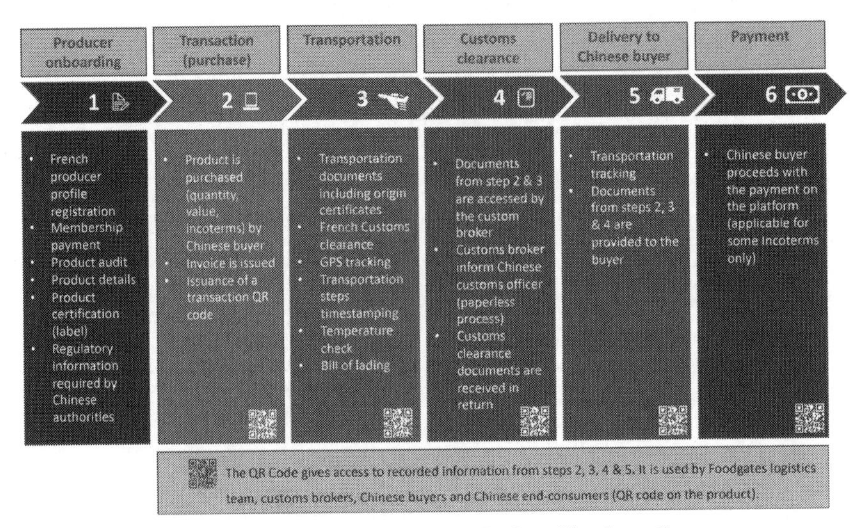

Figure 5. Foodgates flow process. Designed by the authors.

the QR Code as a link, Foodgates's logistics team can monitor the transportation process and inform the customs broker and the Chinese buyers. For the customs clearance process, the customs broker can retrieve information using the QR code and transmit information to the customs officer. Since 2013,[4] China's GACC has generalised a paperless customs clearance process that enables the Customs to directly perform the paperless examination and approval, inspection and release of inbound/outbound goods declared by enterprises via the network by using the information management system of China E-port.[5] Physical inspections of the goods are conducted randomly by the Customs. Paper documents must be submitted for these inspections. Hebe Liu indicated, "Random checks concern between 5 and 20% of the shipments, depending on the nature of the products". After customs clearance is completed, a custom declaration and a CIQ (China Inspection and Quarantine) entry certificate are delivered by the Customs. This documentation is stored on the Foodgates platform. Following the customs clearance, goods are sent to the Chinese buyer. Transportation steps are tracked and stored on the platform up to delivery to the buyer. For certain incoterms, the buyer can also redeem the payment using Foodgates platform.

5.4 *The Added Value of Blockchain*

Foodgates platform's information system

Foodgates platform is a business application. Jonathan Horyn stressed that the "Foodgates platform is a proprietary platform developed for specific needs that is operated single-handedly and is autonomous from the use of blockchain". The Foodgates platform is primarily a B2B marketplace for F&B Asia professionals to buy products from European producers. In addition, the system enables the discovery and trade of products, the organisation of orders, then the tracking and tracing of products from producers to end-consumers, including product information, on the blockchain. The recording on the blockchain is optional and depends on Chinese buyers' choice.

As shown in Figure 6, data from the Foodgates platform are transferred to the blockchain system via BlockFlow technology, a solution developed by Younicorns. BlockFlow enables an interconnection with a large number of blockchains, among others VeChain and BSN. Thus, Foodgates is blockchain agnostic, as it is compatible with many types of blockchain. As described by Jonathan Horyn, "BlockFlow transforms any business process into a blockchain-enabled traceability system".

[4] https://www.china-briefing.com/news/china-deepens-paperless-customs-clearance-pilot-reform/.

[5] http://english.customs.gov.cn/statics/3832116d-eda6-4f36-844e-e076ea9e4113.html.

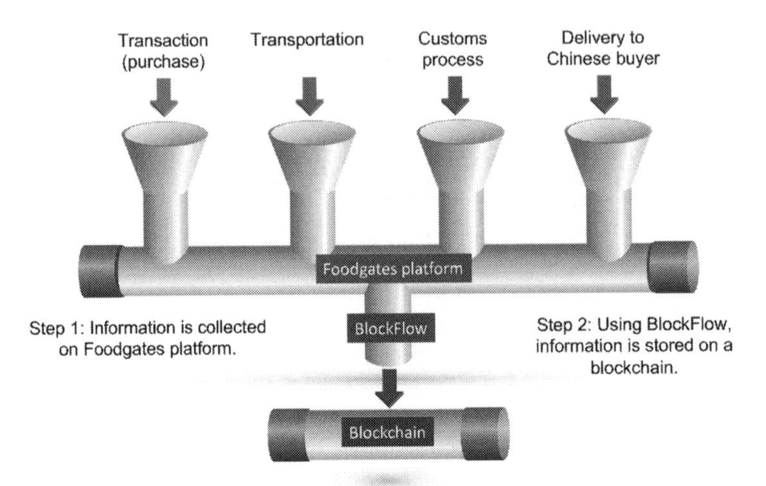

Figure 6. Foodgates data flow to blockchain. Designed by the authors.

From the beginning, Igal Chreky set as a principle to use only public blockchains, and added, "VeChain is a public blockchain which ensures transparency, and it has strong links with China's blockchain ecosystem".

Blockchain transparency

Jonathan Horyn said, "We are recording logistics information on the blockchain for our own needs". Indeed, shipment tracking details such as estimate departure and actual departure, or actual arrival date and time facilitate the overall transportation process monitoring. The transaction QR Code enables a direct link to all pieces of information related to a particular shipment. Hebe Liu added, "Normally, we cannot access the full set of information all at the same place. We need to access the carrier website to check information about the transportation and then we need to check again to ensure the status has not changed. Plus, each shipment might have different brokers, forwarders, and warehouses. Foodgates system enables a transparent tracking of the shipment". Information quality to Chinese buyers is also enhanced.

Besides, on Chinese buyer request, a QR code can be issued and attached to the product. This QR Code enables end-consumers to access information such as product details, logistics traceability through documents timestamped on the blockchain, and documents notarised on the blockchain. Jonathan Horyn explained, "Blockchain enables us to reinforce the validity of information displayed and to reassure Chinese consumers about products' quality". Mathieu Borgé said, "Since the end of 2021, all Chinese buyers request this QR Code service which proves that Foodgates system is reliable and adds value to the transactions".

5.5 Transaction Analysis: Kid's Milk

In August 2020, a Chinese buyer ordered French kid's milk brand Petit Maupassant-Normandy 1905 from a French producer in Normandy, Les Maîtres Laitiers du Cotentin. The following pieces of information were retrieved from Foodgates platform and registered on VeChain blockchain using BlockFlow.[6]

First there is information related to the product that the Chinese buyer accessed before making the decision of placing an order:

- Picture and description of the product
- Labels & certification: Normandy 1905 factory has been certified by the Certification and Accreditation Authority of the People's Republic of China (CNCA). This certification is notarised on the blockchain. This ensures the product is importable in China.
- Map showing places of provenance in Normandy for the farms and the processing factories.
- A video that shows the farmers, the cows and the environment where they live.
- Name of the farmers, size of the farms, number of cows owned and production of milk per year in litres from the farmers that contributed to this production.

The following events, data and documents were made available on the blockchain from the moment the transaction was confirmed by both the Chinese buyer (confirmation to buy the product) and the French producer (confirmation of product availability).

- Dates of milk collection at the farms and reception at the factory in Normandy.
- Recipe elaboration and conditioning dates at the factory.
- Product release date from the factory with a health certificate from French Agriculture Ministry and a certificate of origin from Normandy International Chamber of Commerce (ICC). These certificates are notarised on the blockchain with a timestamp.

The health certificate is an official document specifically issued for export of dairy product to China. This certificate indicates the product's characteristics such as name, description, Ultra-high-temperature processing (UHT), net weight, dates of production and expiration, name and address of the factory where it was produced, shipping location of origin and place of destination, and is signed by an official veterinary

[6] http://demo.foodgates.com/en/milk.

Figure 7. Picture of Petit Maupassant-Normandy 1905 kid's milk cartons.

official under the Ministry of Agriculture. Description of the certificate content is in both the French and Chinese languages.

The origin certificate shows consignor (French milk factory) and consignee (ASI Solutions) names and addresses, country of origin, transport details, description of goods, quantity and net weight. To enable an additional control on the document authenticity, this certificate can be directly retrieved from ICC World Chambers Federation website[7] by using the certificate's number. Description of the certificate content is multilingual, and Chinese is among the displayed languages.

- Dates and timestamp of pick up from the factory, inbound at original warehouse, re-conditioning for transport, outbound at original warehouse, and arrival at airport of origin, Charles De Gaulle airport in Paris. From the moment of pick up, the shipment temperature is monitored, and the results stored on the blockchain up to arrival at final destination. The transportation is organised by ASI Solutions.
- Departure's date and timestamp from airport of origin and bill of lading.
- Arrival's date and timestamp at airport of destination, Shanghai Pudong Airport.
- Dates and timestamp of Customs clearance and departure from airport of destination.
- Dates and timestamp of inbound and outbound at destination's warehouse, and arrival at final destination, the Chinese buyer's place.

Figure 8 highlights the different process steps from the farms in France up to the buyer in China. There are three natures of data collected. The first one is the date at which an event occurred. The second one is the date certified by a timestamp on the blockchain for a particular event. The

[7] https://certificates.iccwbo.org.

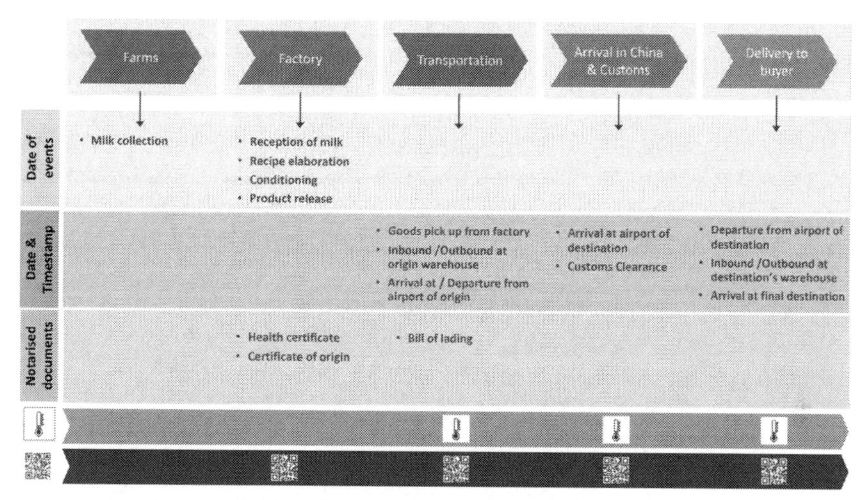

Figure 8. Kid's milk traceability flow. Designed by the authors.

last one is the notarisation of documents on the blockchain. A pictogram at the bottom of Figure 8 indicates that temperature is monitored from the time goods are picked up from the factory up to delivery to buyer's warehouse. Another pictogram indicates that a QR code is printed on the products' packaging which enables access to all information stored on the blockchain once products are released from the factory up to arrival at the buyer's place.

By activating the QR code placed on the products' packaging, all documents are accessible through a webpage which displays descriptions and explanations of the different data in both English and Chinese languages and provides access to notarised documents. Thus, Foodgates platform ensures a complete traceability of the product from the cows to the Chinese consumers' table. Foodgates enables French producers to display the certified quality of their products, and Chinese buyers to promote this quality to their end-consumers. In addition, documents required for the customs clearance process are retrieved from the blockchain to be provided to authorities in China. The proof of provenance, the notarisation of documents, and the tracking of all data with a timestamp on the blockchain support the goods clearance process.

In a standard process, the Chinese Customs broker (the official intermediary) gathers the documents necessary for importing goods and provides them to Chinese Customs agents. Necessary documents vary by product but may include standard documents such as a bill of lading, invoice, shipping list, customs declaration, insurance policy, and sales contract, as well as more specialised documents such as a certificate of origin, a label, a complete description of the products and its production process, an import license, an inspection certificate, and other safety or quality licenses.[8] The customs broker can retrieve from the Foodgates platform all necessary documents using the transaction QR Code as a link to data stored on the blockchain. In addition, the customs broker can access information about the cargo in real-time and therefore prepare in advance, before the cargo reaches China, the online pre-declaration on e-port system. Quality and accuracy of the pre-declaration are important as it ensures a smooth process leading to the issuance of a CIQ entry certificate.

Paperless policy and request for product traceability from Chinese Customs make electronic data information system very relevant to ease customs clearance. "China authorities are moving toward electronic record of traceability", added Jonathan Horyn. Already, as a measure to limit Covid 19 spread risk, Beijing authorities announced in November 2020 that companies transporting or storing imported frozen food items are required to upload details about the products, including their place of origin and import routes, on a government-made online platform.[9] "Papers, transportation timeline and traceability information are to be stored on the official Chinese blockchain BSN", specified Hebe Liu.

5.6 The Use of Blockchain For Traceability

There are a lot of cases that show provenance and traceability of food is a real issue in China. Past scandals, such as the one related to kids' milk [35], highlight how food security is a highly important issue in China. Consumers often do not trust the quality and provenance of products.

Chinese authorities regularly strengthen the regulation to cope with the problem. In December 2021, GACC issued two regulations (Decree 248 and Decree 249) relating to imported food, and affecting overseas food manufacturers, processors, and storage facilities.[10] With effect from January 1, 2022, the new regulations provide for stricter measures regarding Labelling and Packaging, and Supply Chain Safety Trackability. Regarding the latter, foreign food producers are now required to establish a complete and traceable food safety and sanitation control system that ensures products to be exported are produced, processed, and stored in compliance with Chinese laws and regulations (Decree 249, art. 44.).[11]

---

[8] https://www.trade.gov/country-commercial-guides/china-import-requirements-and-documentation.

[9] https://mp.weixin.qq.com/s/nIKhzektF_oAKhvJ3tGp1Q.

[10] https://www.fas.usda.gov/data/china-decrees-248-and-249-january-1-implementation-date.

[11] https://www.morganlewis.com/pubs/2022/01/china-issues-stricter-regulations-on-imported-food#_ftn4.

The use of Blockchain can facilitate compliance to traceability requirements from the authorities. Jonathan Horyn explained, "Blockchain would guarantee the reliability and transparency of information provided". Thus, in 2020, the Guangzhou Municipality's Market Supervision Bureau launched an agricultural product traceability platform based on blockchain to avoid food safety breaches. The solution enables consumers to view traceability information stored on the blockchain.[12]

Several companies operating in China have now developed a blockchain-enabled solution to reassure end-consumers about the quality and provenance of their food products. In 2019, Bright Food, China's second-largest food manufacturer, began to track its dairy products using VeChain. Consumers can trace the provenance of the milk by scanning the QR code on the products. The scan provides information on not only the product's route to shops but also the source of the milk and the type of farm environment conditions it was produced under.[13] Previously, Walmart China had also launched a blockchain food traceability system in the country.[14] In 2021, Australasian Food adopted a blockchain-backed solution to enhance authentication and tracking of beef imported from Australia.[15]

The promulgation of new GACC regulations illustrates that safety throughout the whole food supply chain, from farms to tables, is becoming a high priority of the Chinese authorities and an essential part of the China food regulation enforcement. Beyond specific measures related to Covid 19, China's food import regulation is becoming increasingly complex. As, Igal Chreky commented, "The time given by the authorities to the companies to comply with new regulations is extremely short". For example, GACC issued its Public Notice 2021 number 103 on December 14, 2021, and requested foreign companies to comply by January 1, 2022, when Decrees 248 and 249 took effect.[16] In that respect, the flexibility of Foodgates' information system structure and its blockchain-agnostic nature are an advantage. Should China authorities require to receive current regulatory information throughout a blockchain system, Foodgates would immediately be ready to comply.

[12] https://www.ledgerinsights.com/blockchain-food-traceability-china-food-markets-supervision/.

[13] https://coinrivet.com/china-uses-blockchain-to-track-milk/.

[14] https://www.europeanchamber.com.cn/en/members-news/3303/walmart_china_blockchain_traceability_platform.

[15] https://www.foodnavigator-asia.com/Article/2021/03/02/Year-of-the-ox-Australasian-Food-imports-Argentinian-blockchain-beef-into-China-on-the-back-of-growing-consumption.

[16] https://www.fas.usda.gov/data/china-decrees-248-and-249-january-1-implementation-date.

5.7 Traceability at the Core of China Food Regulation

China food regulation is becoming more and more complex and strict. The new regulations implemented on January 1, 2022, by the GACC are in line with China's updated Food Safety Law that highlights this complexity. In application of Decree 248, all food-producing companies looking to export into China must register with the GACC on a new online platform.[17] This platform called China Import Food Enterprises Registration[18] is part of China E-port data centre. The objective is to collect information before products are exported to China and to qualify foreign producers. Decree 249 focuses on food import safety measures and provide GACC with additional enforcement power to take restriction measures against foreign entities which have violated food safety regulations.[19] The food safety audit system and traceability required by Decree 249 for overseas exporters and food producers is consistent with the latest amendment of China Food Safety Law in 2021. Regulations increase the pressure on foreign actors in an attempt to prevent unsafe products from entering the Chinese market and to be able to quickly react through product recall in case of problems. The traceability before products' arrival in China and after customs clearance up to consumers' table constitutes from now on the core of China food safety enforcement.

We have seen in 5.6 that the Foodgates platform is able to collect an extended traceability flow of information and store it on several blockchains, including BSN, and then make it available to a third party. Thus, the Foodgates platform facilitates China food regulation compliance by providing food safety traceability and speeds up customs clearance process as the pre-declaration can be established by the customs broker early in the process. Besides, the scope of traceability flow of information can be customised to match legal requirements in China. Nevertheless, at this stage, blockchain data are not directly transmitted from Foodgates platform to Chinese Customs. Some recent projects in China that use blockchain-enabled solutions for cross-border customs clearance make plausible the adoption of blockchain by the customs department on a larger scale.

[17] https://www.foodnavigator-asia.com/Article/2021/11/29/Scrutinising-imports-China-launches-mandatory-overseas-food-manufacturer-registration-platform-to-tighten-food-safety#.

[18] https://app.singlewindow.cn/cas/login?_loginAb=1&service=https%3A%2F%2Fcifer.singlewindow.cn%2Fciferwebserver%2Fj_spring_cas_security_check.

[19] https://news.cgtn.com/news/2021-12-29/VHJhbnNjcmlwdDYxNTc5/index.html.

5.8 Adoption of Blockchain for Customs Clearance

Powerbridge Technologies, a China-based technology company specialised in cross-border trade, made public in June 2019 that, as part of the National Customs Cross Border Technology Innovation Initiatives, its Powerbridge Blockchain Cross Border Compliance Platform was adopted by Nanning Customs.[20] Nanning is the capital city of Guangxi province that is located in the south-west of China and shares a nearly 500-mile border with neighbouring Vietnam. Nanning Customs directly manages 26 cross border ports.[21] The objectives were to facilitate customs clearance for cross-border transactions and to improve communication and interoperability between ports and province government authorities.

In October 2019, Chinese authorities officially announced that a blockchain-enabled solution was to be used for cross-border customs clearance in Hangzhou, Zhejiang province.[22] Hangzhou has developed strong logistics and e-commerce industries and deals with more than $37.4 billion of imports and exports annually. Customs authorities claim that the average time for import clearance has been reduced by 85%, from 191 hours to 27 hours, while the export process takes only 12 minutes.[23]

Interestingly, the State Administration of Foreign Exchange (SAFE), a regulator in Chinese Central Government, announced in November 2019 the expansion of its programme launched in March 2019 aiming at using blockchain to support trans-border transactions.[24] This government decision was echoed in 2020 by the launch of BSN blockchain platform. In July 2020, two sub-ecosystems, BSN China for Chinese based projects and BSN International open to any foreign project, were created.[25] The government's apparent objective is to not only to make BSN the standard for blockchain operations in China, but also worldwide through hubs in Hong Kong and Macau [40]. The great emphasis placed by the government on blockchain development lays the foundation for a large adoption in all state and provincial administrations. This strategic move and the creation of BSN are changing the blockchain landscape in China. This new legal and technical framework will greatly promote public and private initiatives. A generalisation of BSN-based e-government applications will

[20] https://www.powerbridge.com/ir/news-content.html?artIdx=32.

[21] https://www.ledgerinsights.com/blockchain-china-customs-powerbridge/.

[22] https://www.ledgerinsights.com/china-blockchain-cross-border-customs-clearance/.

[23] https://www.asiablockchainreview.com/china-adopts-blockchain-to-facilitate-cross-border-trade/.

[24] https://www.cointrust.com/market-news/china-widens-trial-of-its-blockchain-powered-trans-border-funding-platform.

[25] https://forkast.news/chinas-blockchain-infrastructure-project-goes-live-in-hong-kong-macau/.

certainly be extended to the Chinese Customs process. Then, blockchain-enabled information systems will be able to cover not only the traceability requirements but also the customs clearance.

6. Conclusion

The study of Foodgates has illustrated that a blockchain-enabled information system can support food exporters to comply with China food import regulations, either directly by providing a certified tracking of product safety, or indirectly by facilitating the customs clearance process. Greater transparency, instantaneous access to data, and certification of documents make blockchain a RegTech that simplifies companies' compliance with regulations related to import process and supports food safety enforcement.

Foodgates' information system can easily be adjusted to changes in regulations and incorporate more information and documents. This flexibility is important as the food safety regulatory framework needs to adapt to new risk situations and the time to comply with new regulations tends to be short. Besides, the system provides a certified traceability record which is now part of food safety regulation requirements in China. In that respect, Foodgates information system serves the interest of foreign food exporters as well as China policy aiming at ensuring the provision of transparent information to consumers.

With the launch of BSN blockchain, the Central government has been able to push further a blockchain adoption strategy for the government sector. In addition, some provinces have started to adopt blockchain for cross-border transactions clearance. Results show that the clearance process is faster. Besides, blockchain-enabled systems make information and data administration more cost-efficient. Once data are stored on the blockchain, reconciliation of information is no longer required, storage is fully digitised and accessibility to immutable data is permanent. The generalisation of blockchain adoption to enforce food safety regulation seems ineluctable.

APPENDICES

Appendix A. Food-safety regulation history in China from 1949 to 2019.

Stage	Time Span	Administrative System	Main Regulatory Agency	Additional Regulatory Agencies
Hybrid transitional system	1949–1952	Coexistence of industrial self-regulation and governmental supervision	Ministry of Health	N/A
	1953-1978		Related regulatory agencies	Ministry of Health
	1979–1993		Ministry of Health	Regulatory agencies for food producers, regulatory agencies for agriculture, livestock farming and fishing, regulatory agencies for industry and commerce
Comprehensive external regulation	1994–2002	Dominated by external independent regulation	Ministry of Health	General Administration of Quality Supervision, Inspection and Quarantine, State Administration for Industry and Commerce, Ministry of Agriculture
Scientific supervisory system	2003–2007	Comprehensive coordination and segmented regulation	China Food and Drug Administration	General Administration of Quality Supervision, Inspection and Quarantine, Ministry of Health, Ministry of Agriculture, State Administration for Industry and Commerce
	2008–2010		Ministry of Health	General Administration of Quality Supervision, Inspection and Quarantine, State Administration for Industry and Commerce, Ministry of Agriculture, China Food and Drug Administration
	2011–2012		Food Safety Committee	General Administration of Quality Supervision, Inspection and Quarantine, State Administration for Industry and Commerce, Ministry of Agriculture, China Food and Drug Administration, Ministry of Health
Modernization of governance system	2013–2018	Trinary Regulatory System	Food Safety Committee	China Food and Drug Administration, State Administration for Industry and Commerce, General Administration of Quality Supervision, Inspection and Quarantine, Ministry of Agriculture
	2018–2019		State Administration for Market Regulation	Ministry of Agriculture and Rural Affairs

Adapted from Gao et al. [16]. Sources: Hu [23] and Liu et al. [29].

Appendix B. China government's evolution from food hygiene to food safety regulation.

Food hygiene management			**Food safety governance**	
I	II	III	IV	V
1949–1979	1979–1995	1995–2009	2009–2015	2015–2022
Centralised Management	Multisector Management	Matrix Management	Process Management	Integrated Management
Ministry of Health; Departments of five ministries involved	Departments of agriculture, forestry, animal husbandry, aquaculture, grain, supply and marketing, business, light industry, and trade.	Institutions of the Ministry of Health at all levels over the country; Departments of local government in the jurisdiction.	Food Safety Committee for overall guidance; Relevant Ministries for process management	Food Safety Committee for overall guidance; CFDA (SAMR, 2018) for safety supervision; NHFPC (NHC, 2018) for risk monitoring, risk assessment, and safety standard.
Regulations on the Management of Food Hygiene (Trial Implementation) 1964	Food Hygiene Law (Trial Implementation) 1982	Food Hygiene Law 1996	Food Safety Law 2009	Food Safety Law (Amended) 2015, 2018 and 2021 (29 April).

Adapted and updated from Liu et al. [29].

Appendix C. Food import china process flow chart. Source: www.rjs.cn

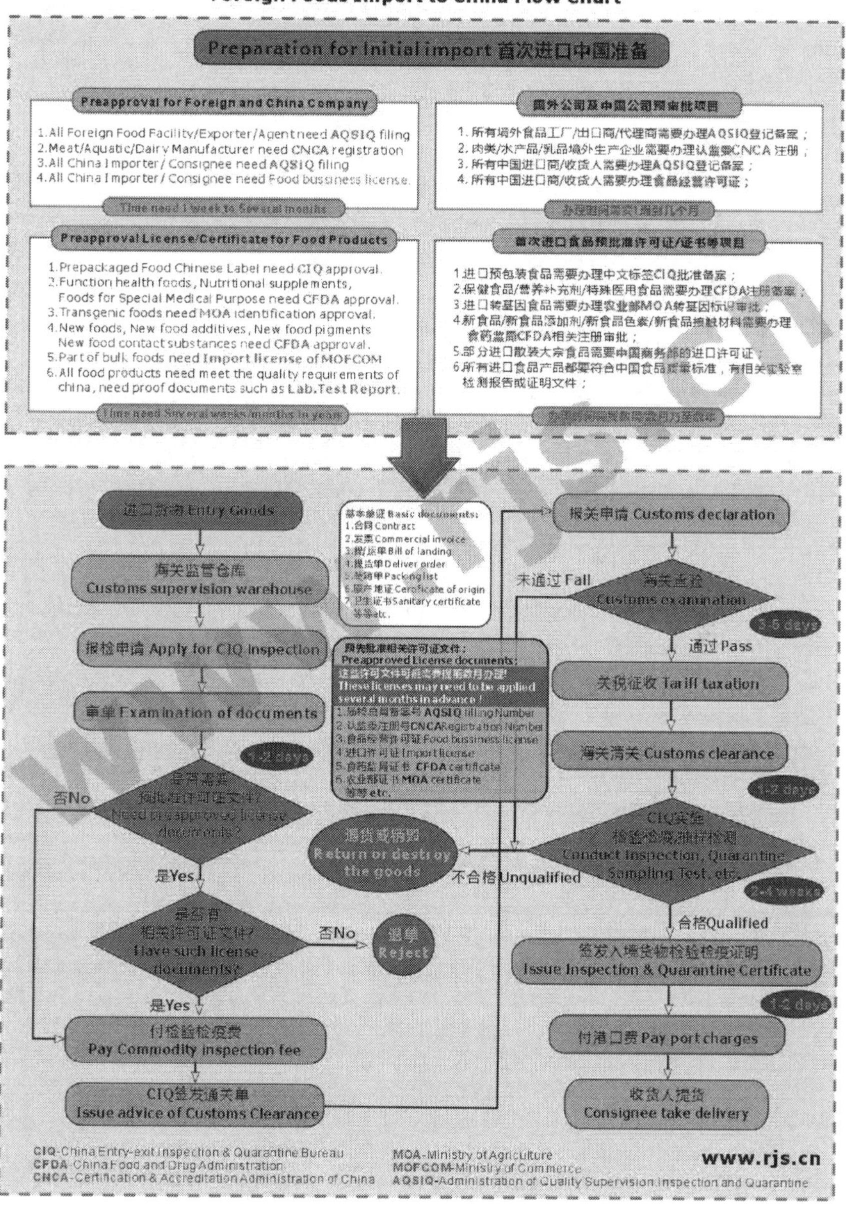

Appendix D. Foodgates stakeholders' overview

ASI Group was created in 2008 in Shanghai by Mr. Dov Chreky, the company Founder and Chairman. Today, ASI Group counts three major subsidiaries, ASI Logistics, ASI Solutions and ASI Movers. Located in six countries in Europe and Asia, ASI Group has more than 200 employees worldwide. The Foodgates project has originated from the initiative of ASI Group that selected the different partner companies; see https://group-asi.com

Younicorns is a corporate studio, co-creating successful start-ups with corporates. They turn great ideas into real products. Younicorns has technologically developed the Foodgates platform. Younicorns was selected by ASI Group because of its past experience in managing projects between Europe and China. Younicorns has also developed the technology BlockFlow that connects business processes to blockchain solutions in the supply chain field. https://younicorns.io - https://block-flow.com

Founded in 2015, **VeChain** Technology is one of the earliest blockchain technology companies in the world and provides blockchain-enabled solutions suited to business needs. With international offices in China, Europe, US, Singapore, and Japan and relying on strategic partners, PwC and DNV, VeChain has established partnerships with many leading enterprises in various industries, including Walmart China, Bayer China, BMW Group, LVMH, etc. VeChain was selected by ASI Group because of its involvement in several China-based projects including partnership with some local governments. Foodgates is part of the VeChain partnerships' list.[26] https://www.vechain.com

DNV is the world's leading certificatory with a strong expertise in assurance and risk management. Founded in 1864 in Norway, DNV is a recognized advisor for the maritime industry. In the Foodgates project, DNV acts as a certifier for origin and labels of products. https://www.dnv.com

BSN, Blockchain-based Service Network, is a cross-cloud, cross-portal, cross-framework global infrastructure network used to deploy and operate all types of blockchain applications. The State Information Centre of China, a government agency, is one of the six entities that established the BSN Development Association. The other five entities are leading Chinese companies in the telecommunications industry (China Mobile Communications Corporation Design Institute Co., Ltd. And China Mobile Communications Corporation Government and Enterprise Service Company), financial industry (China UnionPay Corporation and

[26] https://vechaininsider.com/partnerships/a-complete-list-of-vechain-partnerships/#asi_group.

China Mobile Financial Technology Co., Ltd.), and software industry (Beijing Red Date Technology Co., Ltd.).[27] BSN is referred to as the official "China" blockchain. In January 2022, China picked 15 city governments and 164 other entities including companies, universities, hospitals, and industry groups to carry out a trial of BSN.[28] The use of BSN is foreseen to be the key element of the e-government strategy of the central authorities. Foodgates is compatible with BSN. https://www.bsnbase.com/

Glossary

Regulatory technology: This term emerged following the 2008 global financial crisis and the considerable reinforcement of the legislation. The aim was to use technologies to ease the burden of regulatory compliance for financial institutions.

Blockchain-based RegTech system: Blockchain is considered to be a regulatory technology, especially in the finance field. The potential application of an information system enabled by blockchain to facilitate food safety compliance and enforcement is a step toward a larger adoption of blockchain as a RegTech.

Food safety enforcement: Food safety is a public health concern and a prerogative of the governments that act as regulators. Rules and regulations must be complied with by all actors aiming at selling, importing, or exporting products in a particular territory.

Acknowledgements

We thank all interviewees, Igal Chreky, Hebe Liu, Jonathan Horyn, Mathieu Borgé and Jérôme Guillères, from ASI Group, Foodgates and VeChain for providing the information that supported this research.

References

[1] Arner, D. W., Barberis, J. and Buckey, R. P. 2016. FinTech, RegTech, and the reconceptualization of financial regulation. Northwest. *J. Int. Law Bus.*, 37: 371.

[2] Baxter, P. and Jack, S. 2008. Qualitative case study methodology: Study design and implementation for novice researchers. *The Qualitative Report*, 13(4): 544–559.

[27] BSN Introductory White Paper: https://bsnbase.io/g/main/documentation.

[28] https://www.bloomberg.com/news/articles/2022-01-30/china-picks-cities-entities-to-take-part-in-blockchain-trials.

[3] Beck, R., Stemi, J. C., Lollike, N. and Malone, S. 2016. Blockchain - The gateway to trust-free cryptographic transactions. *Proc. ECIS Turkey*, 1–14.
[4] Beck, R., Avital, M., Rossi, M. and Thatcher, J. 2017. Blockchain technology in business and information systems research. *BISE*, 596: 381–384.
[5] Behnke, K. and Janssen, M. F. W. H. A. 2020. Boundary conditions for traceability in food supply chains using blockchain technology. *Int. J. Inf. Manage*, 52: 101969.
[6] Cai, L., Sun, Y., Zheng, Z., Xiao, J. and Qiu, W. 2021. Blockchain in China. Commun. ACM, 64(11): 88–93.
[7] Calvin, L., Gale Jr., H. F., Hu, D. and Lohmar, B. 2006. Food safety improvements underway in China. *Economic Research Service/USDA*, 4(6): 16–21.
[8] Chammem, N., Issaoui, M., Dâmaso De Almeida, A. I. and Delgado, A. M. 2018. Food crises and food safety incidents in European union, United States, and Maghreb area: Current risk communication strategies and new approaches. *J. AOAC Int.*, 101(4): 923–938.
[9] Chu, M. 2021. The limits to the internationalisation of regulation: divergent enforcement strategies in China's food safety regulation. *Policy & Politics*, 1–21.
[10] Creswell, J. W. 2007. Qualitative inquiry & research design: choosing among five approaches (2nd ed.). Thousand Oaks: Sage Publications.
[11] Crosby, M., Pattanayak, P., Verma, S. and Kalyanaraman, V. 2016. Blockchain technology: Beyond bitcoin. *Applied Innovation Review*, 2: 6–10.
[12] De Filippi, P. and Hassan, S. 2018. Blockchain technology as a regulatory technology: From code is law to law is code. arXiv preprint arXiv:1801.02507.
[13] Dong, L. 2018. What's the Future of Blockchain in China? World Economic Forum. https://bit.ly/36Sll7u (accessed 30 October 2021).
[14] Francisco, K. and Swanson, D. 2018. The supply chain has no clothes: Technology adoption of blockchain for supply chain transparency. Logistics, 2(1): 2.
[15] Gale, F. 2021. China's Refusal of Food Imports. ERR-286, U.S. Department of Agriculture, Economic Research Service.
[16] Gao, Q., Huang, Y., Liang, Q., Sui, Y. and Zheng, Y. 2020. Food Safety Regulatory Enforcement in China: A Data-Driven Approach. SSRN: http://dx.doi.org/10.2139/ssrn.3658748.
[17] Gozman, D., Liebenau, J. and Aste, T. 2020. A case study of using blockchain technology in regulatory technology. *MIS Quarterly Executive*, 19(1): 19–37.
[18] Guo, Z., Bai, L. and Gong, S. 2019. Government regulations and voluntary certifications in food safety in China: A review. *Trends in Food Science & Technology*, 90: 160–165.
[19] Hawlitschek, F., Notheisen, B. and Teubner, T. 2018. The limits of trust-free systems: A literature review on blockchain technology and trust in the sharing economy. *Electron. Commer. Res. Appl.*, 29: 50–63.
[20] Henson, S. and Heasman, M. 1998. Food safety regulation and the firm: Understanding the compliance process. *Food Policy*, 23(1): 9–23.
[21] Hou, H. 2017. The application of blockchain technology in E-government in China. Proc. *ICCCN Canada*, 1–4.
[22] Hou, J., Wu, L. and Hou, B. 2020. Risk attitude, contract arrangements and enforcement in food safety governance: a China's agri-food supply chain scenario. *Int. J. Environ. Res. Public Health*, 17(8): 2733.
[23] Hu, Y. 2016. Chinese policy on the evolution of food safety: Concept and practice. *Reform*, (5), 25–40.
[24] Huang, J. and Yang, G. 2017. Understanding recent challenges and new food policy in China. *Global Food Security*, 12: 119–126.
[25] Hughes, L., Dwivedi, Y., Misra, S., Rana, N., Raghavan, V. and Akella, V. 2019. Blockchain Research, Practice and Policy: Applications, Benefits, Limitations, Emerging Research Themes and Research Agenda. *Int. J. Inf. Manage.*, 49: 114–129.

[26] Kendall, H., Naughton, P., Kuznesof, S., Raley, M., Dean, M., Clark, B., Stolz, H., Home, R., Chan, M.Y., Zhong, Q., Brereton, P. and Frewer, L. J. 2018. Food fraud and the perceived integrity of European food imports into China. *PloS one*, 13(5): e0195817.

[27] Lacity, M. C. 2018. Addressing key challenges to making enterprise blockchain applications a reality. *MISQS*, 17(3): 201–222.

[28] Lemieux, V. 2016. Blockchain for Recordkeeping; Help or Hype? Social Sciences and Humanities Research Council of Canada Knowledge Synthesis Report, October. https://doi.org/10.13140/RG.2.2.28447.56488.

[29] Liu, Z., Mutukumira, A. and Chen, H. 2019. Food safety governance in China: From supervision to coregulation. *Food Sci. Nutr.*, 7(12): 4127–4139.

[30] Lootsma, Y. 2017. Blockchain as the newest regtech application—the opportunity to reduce the burden of KYC for financial institutions. *Banking & Financial Services Policy Report*, 36(8): 16–21.

[31] Micheler, E. and Whaley, A. 2020. Regulatory technology: replacing law with computer code. *Eur. Bus. Organ. Law Rev*, 21(2): 349–377.

[32] Ogus, A. 1992. Information, error costs and regulation. *Int. Rev. Law Econ.*, 12(4): 411–421.

[33] Ogus, A. 1994. Regulation: Legal form and economic theory (Clarendon law series). Oxford: Clarendon Press.

[34] Parra Moyano, J. and Ross, O. 2017. KYC Optimization Using Distributed Ledger Technology. *Bus. Inf. Syst. Eng.*, 59(6): 411–423.

[35] Pei, X., Tandon, A., Alldrick, A., Giorgi, L., Huang, W. and Yang, R. 2011. The China melamine milk scandal and its implications for food safety regulation. *Food Policy*, 36(3): 412–420.

[36] Perboli, G., Musso, S. and Rosano, M. 2018. Blockchain in logistics and supply chain: A lean approach for designing real world use cases. *IEEE Access*, 6: 62018–62028.

[37] Qian, J., Wu, W., Yu, Q., Ruiz-Garcia, L., Xiang, Y., Jiang, L., Shi, Y., Duan, Y. and Yang, P. 2020. Filling the trust gap of food safety in food trade between the EU and China: An interconnected conceptual traceability framework based on blockchain. *Food Energy Secur.*, 9(4): e249.

[38] Saberi, S., Kouhizadeh, M., Sarkis, J. and Shen, L. 2019. Blockchain technology and its relationships to sustainable supply chain management. *Int. J. Prod. Res.*, 57(7): 2117–2135.

[39] San Juan, I. H. 2020. The Blockchain Technology and the Regulation of Traceability: The Digitization of Food Quality and Safety. European Food and Feed Law Review. 15(6): 563-570.

[40] Stockton, N. 2020. China takes blockchain national: The state-sponsored platform will launch in 100 cities. *IEEE Spectrum.*, 57(4): 11–12.

[41] The People's Government of Beijing Municipality 2020. The blue book on the innovative blockchain applications in the field of government services in Beijing (1st ed.). Beijing, China.

[42] Tian, F. 2016. An agri-food supply chain traceability system for China based on RFID & blockchain technology. *Proc. ICSSSM China, IEEE*, 1–6.

[43] Treiblmaier, H. 2019. Toward more rigorous blockchain research: Recommendations for writing blockchain case studies. *Frontiers in Blockchain*, 2: 1–15.

[44] Wang, Y., Han, J. and Beynon-Davies, P. 2019. Understanding blockchain technology for future supply chains: A systematic literature review and research agenda. *Supply Chain Manag.*, 24(1): 62–84.

[45] Xiu, C. and Klein, K. K. 2010. Melamine in milk products in China: Examining the factors that led to deliberate use of the contaminant. *Food Policy*, 35(5): 463–470.

[46] Yin, R. K. 2018. Case study research: Design and methods (6th ed.). Los Angeles, California: Sage Publications.

4

Transforming Trade Finance via Blockchain

The We.Trade Platform

David Petersen

1. Introduction

Of the many sectors of supply chain management which offer the potential for digital transformation via blockchain technology, trade finance represents one of the most tangible and the most valuable use cases. Theory and anecdotal evidence suggest that blockchain has the potential to transform key elements of the trade finance sector, addressing needs for more transparent and reliable processes, greater control over risks, and for the robustness of processes in the face of supply chain disruptions such as COVID-19.

This chapter examines these transformative capabilities of blockchain through their real-world application in the we.trade digital platform for trade finance. Specifically, this case study of we.trade will assess how the use of blockchain can enable the members of a digital platform to optimize their inter-organizational processes, and as a result, generate competitive advantages.

University of Strasbourg, 61 avenue de la Forêt-Noire, 67000 Strasbourg, France.
Email: david.petersen@em-strasbourg.eu

2. Background

2.1 Trade Finance

It is estimated that up to 80%, or approximately USD 16 trillion, of merchandise trade, depends upon trade finance in some form [1]. Buyers and sellers may use a wide range of instruments and terms to facilitate trade, including loans, guarantees, and the factoring or financing of receivables and inventories. They may engage various intermediaries such as banks, financial institutions and service providers to do so [1].

Trade can typically be categorized as being either documentary (via letters of credit), or open account (payment after delivery), in nature. Although the letter of credit instrument provides greater control over the process, these are costly and time consuming to execute. For this reason, open account trade has been growing at a faster rate worldwide [1]. However, the use of open account trade makes more challenging the optimization of cash flow and the mitigation of the risk of payment not being received and strains the resources of small-and-medium-sized enterprises (SMEs) in particular. Combined with the lack of visibility and the possibility of fraud due to paper-based processes, and the varying degrees of trust present between trading partners, there is a great need for reliable trade and trade finance processes.

Gaining access to trade finance can be a major issue at any time for SMEs, with approximately 65 million SMEs being credit constrained and rejection rates for credit applications running at 40% [2]. In times of international crisis, such as the COVID-19 pandemic, the need for trade finance can become critical. Due to the greater perceived risks, the availability of credit is reduced while the cost of credit increases. In particular, less open account credit is forthcoming from trading partners and less financing available from financial intermediaries [3], while the necessary trade credit insurance coverage becomes more expensive and more difficult to secure [4].

2.2 Blockchain in a Business Context

Blockchain technology is comprised of several unique components, and the most important of these from a business perspective maybe its decentralized design, the enhanced security it provides, and the 'smart contracts' that it enables. Blockchain allows transactions to be submitted and validated by appointed nodes across a distributed database and network. New data is appended to previous data such that the chain of data is practically immutable (that is, any attempted tampering would be detected), and all data is secured via encryption [5,6]. Smart contracts can automate the processing of transactions upon being triggered by predefined criteria.

The vast majority of business blockchains are permissioned (that is, private) in nature. In this design, an authority representing the group or consortium which operates the blockchain will control which entities can join, transact, and endorse (that is, validate) transactions on the blockchain. Typically, this will include digital management of the identity of the members, and off-blockchain contracts detailing rights and responsibilities. This is very different from those permissionless (that is, public) blockchains such as Bitcoin, where is no central controlling body and where the network manages itself independently. While the nodes allowed to submit transactions and those responsible for endorsing transactions on a permissioned blockchain are appointed by the controlling authority, on a permissionless blockchain any node can submit transactions and any node can endorse the transaction if it 'wins' the right to do so (for example, by 'mining' on Bitcoin).

2.3 Business Networks as Digital Platforms

A digital platform can be thought of as a two-sided or multi-sided market which electronically connects buyers and sellers and intermediaries within a network [7]. As the number of buyers and sellers on the platform increases, the ecosystem grows in usefulness and value for its members. This is the network effect [8], which represents a virtuous circle of ever-extending mutual benefits for members [9].

Digitization enables processes to be performed independent of location, removes the need for the holding of physical assets, and allows the creation of a powerful platform ecosystem within which supplementary service providers can participate and add value [10,11]. Digital platforms can therefore provide a structure through which participants can seek to optimize their inter-organizational connections.

With the relational view of Dyer and Singh [12] this can be taken further. The relational view—rather than focus on industry forces [13], or on firm-specific factors [14,15]—extends the area of analysis into the business networks within which the organization participates, and suggests that optimizing their inter-organizational processes can represent a method by which organizations can generate competitive advantages.

2.4 Blockchain Enabled Digital Platforms

How can a blockchain enabled digital platform achieve the objective of optimizing inter-organizational processes and generating competitive advantages for its members? It is suggested that a structure comprising the following elements acting in concert is required [16]:

1. At an administrative level, there must be an acceptable and impartial set of rules governing the platform and member behavior.
2. At the infrastructure level, the blockchain must be technically secure and reliably governed.
3. At the application level, the blockchain must use smart contracts to automatically process transactions, and to govern adherence to the trade terms and conditions agreed between the parties.

we.trade is one of the relatively few digital platforms in operation which utilizes blockchain technology to provide a complete end-to-end ecosystem, in this case with the purpose of facilitating trade transactions and trade finance. This chapter will examine we.trade according to the criteria outlined above.

3. we.trade—A Blockchain Trade Finance Platform

3.1 Development

The origins of we.trade date from 2015, when the then-head of trade finance at Belgian bank KBC, Hubert Benoot, upon being introduced to blockchain technology by the KBC IT department, recalled an event from two to three years before that. At a round table held for SME customers, a potato farmer had been asked why he didn't export his produce outside of Belgium. The farmer had responded that it was too risky: when he didn't know a potential new customer, how could he trust that partner, and how could he be sure that he would get paid? [17,18].

Benoot connected the need for providing trust throughout the trade process with the new blockchain technology, and the journey was begun. Several European banks were approached with a suggestion of collaboration, and a digital platform was envisioned as an end-to-end ecosystem to connect buyers, sellers, and service providers (especially the banks involved, but also including insurers, logistics organizations, and other entities). The platform would simplify the trading process, reduce risk by integrating participants and automating transactions, and enable the banks to offer any required financial services [19].

From the perspective of the banks, we.trade would allow them to sell new financial products to the previously under-serviced market of companies engaged in open-account trading, without compromising their existing profitable letter of credit business. It has been estimated by we.trade that over 95% of intra-Europe trade is based upon open-account processes, and that this rises to 99% when considering only SMEs [20]. While expanding the volume of trade facilitated, the we.trade platform would also eliminate much of the risk inherent in open-account trade for all parties.

The banks agreed to cooperate on a proof of concept called the Digital Trade Chain, which was completed in 2016. A consortium of shareholder banks was formally established with seven initial members in 2017, with two additional banks joining later in 2017, and with a further three banks joining in 2019 upon the folding of the Batavia trade finance consortium. Meanwhile, the next generation of we.trade platform was designed. While the model can be said to be bank-centric in that the platform was led and funded by a consortium of banks, the needs of the buyers and sellers were encompassed in the design, and service providers were intended to play an essential part. Indeed, both IBM and CRIF (a credit information provider) have subsequently also become we.trade shareholders. The participants of the we.trade platform are presented in Figure 1.

The we.trade joint venture (JV) company was registered in Ireland in 2018, a location which was considered certain to remain within the EU, and which offered a favorable environment for IT startups. A beta version of we.trade was produced in 2018, and the we.trade platform was formally launched in 2019. The timeline for the development of we.trade is shown in Table 1.

As of January 2022, the we.trade platform provides extensive coverage across Europe, including the shareholder banks Caixa (Spain), Deutsche

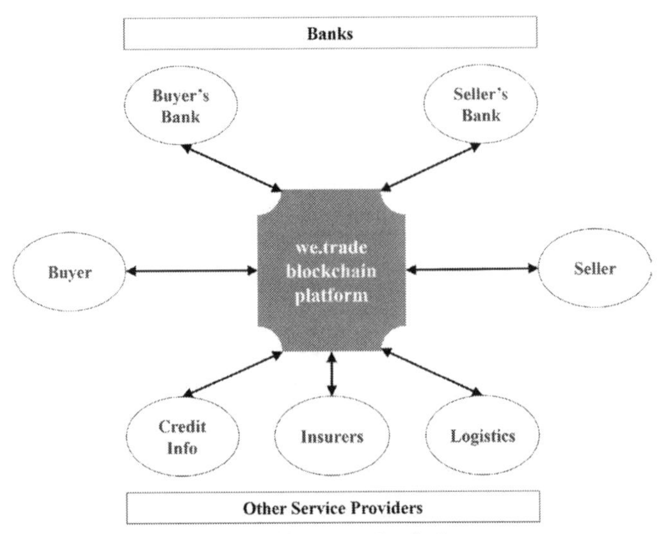

Figure 1. The we.trade platform.

Table 1. we.trade timeline.

2015		Idea formed
2016		Proof of concept is created called "Digital Trade Chain"
2017	Q1	we.trade consortium is formed with initial members
		(Deutsche Bank, HSBC, KBC, Natixis, Rabobank, Société Générale, UniCredit)
	Q4	More banks join the we.trade consortium (Nordea, Santander)
2018	Q2	Rulebook is established to provide governance
	Q3	Joint venture company is established in Ireland
	Q4	First live transaction is processed
2019	Q1	Commercial launch of we.trade
	Q1	Batavia consortium folds and its member banks join we.trade (Caxia, Erste, UBS)
2020	Q2	IBM becomes a shareholder
	Q3	Additional banks become licensees (not shareholders)
	Q4	Capital is injected by shareholders
2021	Q1	CRIF (credit information provider) becomes a shareholder
	Q1	Major platform upgrade is released
	Q1	First phase of ERP integration is released
	Q4	First non-EU bank joins we.trade (Akbank of Turkey)

Bank (Germany), Erste (Austria), HSBC (UK), KBC (Belgium), Natixis (France), Nordea (Scandinavia), Rabobank (Netherlands), Santander (Spain), Société Générale (France), UBS (Switzerland), and UniCredit (Italy), and the non-shareholder banks Akbank (Turkey), CBC (Belgium), Eurobank (Greece), UniCredit (Germany), and ČSOB, Komerční Banka and Česká Spořitelna (Czech Republic). To date, agreement has been maintained between the shareholder banks and non-shareholder banks over the benefits which may accrue from shareholding. This might be as expected at the current stage in the development of we.trade, at which it is requiring continued investment, rather than providing returns. This accord between stakeholders may be more challenging to achieve in future stages if and when the investment/return balance changes. The banks which participate on the platform contribute a license fee for the right to do so.

The consortium which controls we.trade today is comprised of its shareholder banks, and shareholder service providers IBM and CRIF. Together these organizations own the shares of the we.trade JV, and nominate executives to its board of directors. As of the start of 2022, the chairperson of the board of we.trade is Agnès Joly, who is also Head of Innovation and Strategy - Global Transaction Banking for Société Générale.

Table 2. we.trade Volume as of the end of 2021.

Metric	Measure
Number of Participating Companies	400
Number of Transactions Performed to Date	1,500
Value of Transactions Performed to Date	EUR 120 million
Increase in Transactions 2021 over 2020	104%

The role of the we.trade consortium is to set and to monitor the direction and rules of the platform, while the role of the we.trade JV is to provide and operate the platform itself. Communication with the trading companies who buy and sell on the platform is performed by the participating banks, rather than by the we.trade JV. Currently, the we.trade JV does not have a CEO or General Manager, and is led by its CTO, Mark Cudden [21].

Since its commercial launch in 2019, we.trade has steadily increased its membership. At the end of 2021, we.trade published the transaction volumes [22] presented in Table 2.

3.2 Trade Finance on we.trade

While a large number of digital platforms have been created to address the needs of the trade finance sector, these differ in their approaches. As of 2021 for example, Contour and Komgo focus on the letter of credit process, Marco Polo on corporate finance, and eTradeConnect upon documentation flows [23].

The we.trade platform is directed to open account trade. While we.trade does not exclusively target SMEs, given the limited resources of this type of organization, the SME has the greatest need for trade finance solutions in open account trade. The major steps of this process when conducted on we.trade [24] are summarized in Figure 2.

While the standard trading steps of purchase order creation (step 1), seller shipment (step 3), and buyer receiving (step 4) certainly exist in this process, the unique aspects of we.trade can be seen in the steps for the optional new banking products of Bank Payment Undertaking (BPU) (step 2) and BPU Financing (step 5), and the auto-settlement via smart contract of the payment to be made by the buyer's bank (step 6). These key elements on the we.trade process for trade finance are further described below [25].

Bank Payment Undertaking (BPU): Typically, at the seller's request, the buyer can apply for a BPU from their bank, and if their bank accepts, then that bank will guarantee to make the required payment directly to the seller's bank, upon the satisfaction of the settlement conditions as

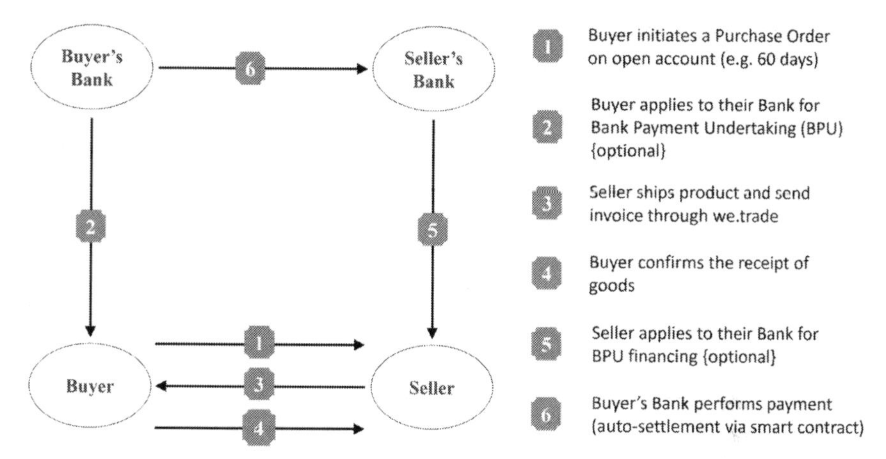

Figure 2. The we.trade process.

determined by a smart contract. Thus, when the buyer arranges a BPU, the seller is guaranteed to receive payment once the settlement conditions are met. In return for this lower risk, the seller may grant the buyer preferential terms, such as longer payment terms, which would in turn help their cash flow.

The BPU is the we.trade extension of the Bank Payment Obligation (BPO) standard for the exchange of electronic documentation which was introduced by the International Chamber of Commerce in 2013 [23], with we.trade introducing the capability of automated settlement of the payment process enabled via smart contract.

BPU Financing: If the seller requires additional cash flow support within the period before receiving the payment from the buyer, the seller can apply to their bank for financing, using the BPU as a negotiable instrument.

Auto-Settlement: If a smart contract determines that a predefined settlement condition has been met (e.g., the goods have been delivered), then the buyer's bank will process the payment to the seller's bank. This payment may be fully automated if agreed in advance by all parties.

The trade finance process on we.trade is highly automated and is based upon event triggers controlled by smart contracts. Upon any of the steps in the we.trade process occurring, a notification is sent to each involved party in real time, and the data is written to the we.trade blockchain.

3.3 Buyers & Sellers in we.trade

According to a breakdown of we.trade members by sector in 2020, 46% of members were from industrial/manufacturing, 40% from clothing and apparel, and 5% from foods [26]. Published reports have referenced the companies listed in Table 3 as we.trade members [19,26–30].

An illustrative member of the we.trade platform is Ekoï, a French company providing clothing and accessories for cyclists worldwide. According to an interview with Ekoï management in late 2021, they use we.trade for all transactions with four of their Italian suppliers who are also on we.trade, and had completed over 50 transactions in the prior two months. Since the order cycle for their products is relatively long, and the quantities large, Ekoï values the we.trade platform for facilitating communication between themselves and the sellers at each stage of the process, together with the fact that all information is assured to be complete and transparent. Ekoï management stated that they will ask all new suppliers to join we.trade if possible.

Ekoï financial management also expressed appreciation for the financial products available on we.trade as being much simpler and less expensive than letters of credit. The use of a BPU allows sellers to extend payment terms, providing time for the buyer to sell most of the delivered products in that period. The sellers meanwhile could request BPU Financing as required and receive confirmation of that facility almost immediately.

Table 3. we.trade Members.

Member	Country	Industry	Bank
Actherm	Czech Republic	Steel	CSOB
Ekoï	France	Durables	Société Générale
Flattered	Sweden	Footwear	Nordea
Fluid Pumps	UK	Industrial	HSBC
Polimer Tecnic	Spain	Plastics	Caixa

3.4 How Blockchain Enables The we.trade Platform

This section examines how blockchain enables the we.trade platform via a structure comprising three levels of governance: application level mechanisms which specify the rules by which the network and its members interact, infrastructure level mechanisms which provide the core blockchain functionality, and application level mechanisms by which smart contracts automate processing and enforcement [16].

3.4.1 Administrative Level Mechanisms

It has been proposed that for a platform to be effectively enabled by a blockchain-driven governance structure, at the administrative level there must be an acceptable and impartial set of rules governing the network and member behavior in order for members to be willing to participate [16]. Can this be said to be present at we.trade?

Before being commercially launched in January 2019, the we.trade consortium designed and implemented a common framework to govern platform operations, which it called its rulebook. The rulebook was based upon English trade laws [27], perhaps reflecting the significant role of HSBC in its creation [31].

Each of the banks, buyers and sellers using we.trade must commit to adhere to the provisions of the rulebook [25], which include the specification of:

- the rights and responsibilities of the parties
- the enrollment processes for a bank, buyer, or seller
- the operation of platform services, and handling of exceptions
- the operation of bank products, and handling of exceptions
- security, confidentiality, data protection, and privacy
- termination of access of a buyer, seller, or bank
- dispute resolution processes, between buyer and seller, or with a bank, or between banks

As per the rulebook, new banks can be added to we.trade with the agreement of the board representing the consortium. New buyers and sellers are introduced to the platform by one of the participating banks, and that bank is then responsible for performing the know-your-customer (KYC) processes required to verify the company's bona-fides.

Perhaps most importantly of all the provisions of the rulebook, each member explicitly agrees to accept the results of the processing of any smart contract created on the we.trade platform [25].

The general manager of we.trade in the year following the creation of the rulebook, Ciaran McGowan, was quoted as saying that establishing the governance rulebook took more time than the technical building of the platform itself [26].

With the board of the we.trade JV comprised of representatives from across the shareholder companies, no single shareholder has a significantly larger say than another, and this has helped to avert suggestions of impartiality between the banks [21]. Furthermore, the rulebook specifies that all participants will have equal rights and responsibilities [25].

IBM's representative involved in the we.trade project, Parm Sangha, mentioned in an interview and in the press that, in general, banks prefer a platform run by a neutral entity, rather than join one dominated by another bank, in order to avoid the possibility of bias in the use of platform assets and in the value derived from those assets [18,29]. Neither do the most recent additions to the shareholding consortium, IBM and CRIF, hold any greater control than any of the other consortium members.

From interviews with stakeholders and a review of relevant media, there does indeed appear to be an acceptable and impartial set of rules governing the we.trade platform and its members.

3.4.2 Infrastructure Level Mechanisms

It has been proposed that for a platform to be effectively enabled by a blockchain-driven governance structure, at the infrastructure level members must be assured that the technology is secure and reliably governed in order to place their trust in it [16]. Can this be said to be the case at we.trade?

we.trade utilizes the Linux Foundation's Hyperledger Fabric as its blockchain technology, with the infrastructure and initial solution implemented by IBM on its blockchain platform. The selection of Hyperledger and IBM took the we.trade consortium over six months, and was greatly influenced by the stability both of Hyperledger (already an established technology at that time) and of IBM (which was expected to continue in the blockchain business line for the long term). Further extension, testing, and support has also been performed by we.trade JV technical staff.

The we.trade web interface layer is deployed as a SaaS (Software as a Service) public cloud service from several European data centers, with the additional option for a bank to have an instance installed on a data center in any location, for example if so required by local country regulations [32].

For blockchain functions, we.trade uses the 'channels' feature in Hyperledger Fabric to form a private network within which transactions are only shared with the members of that channel. Each bank is a node on the network. Data sent to a channel is only present on authorized nodes, and nodes only store the data of the channels to which they are authorized. The detailed data for a trade transaction is uploaded to a channel used only by the buyer, the seller, and their respective banks. Track and trace data can be kept anonymously for a wider set of involved parties. The blockchain does not contain personal data, and neither does it hold document attachments directly, rather it stores a reference to external file storage. It is noted that this channels design is often used in permissioned business oriented blockchain networks which utilize Hyperledger Fabric, for example in the Tradelens network for the management of sea transport [33].

Most importantly, each member of we.trade has confirmed with their acceptance of the rulebook that transactions created on the we.trade platform are fully enforceable, and that communications sent through the we.trade platform are equivalent to those in written form [25].

It does appear that the we.trade platform is technically secure and reliable, and there has been no mention in interviews with stakeholders or in the media of any incident suggesting the contrary.

3.4.3 Application Level Mechanisms

It has been proposed that for a platform to be effectively enabled by a blockchain-driven governance structure, at the application level smart contracts must automate the processing of transactions and must govern adherence to the trade terms and conditions agreed between the parties [16]. Can this be said to be the case at we.trade?

In contrast to the physical document checks which drive the letter of credit process, transactions on we.trade are processed automatically through smart contracts. During the initial we.trade implementation, each of the consortium banks provided their business requirements, and these were standardized into a form that could be further rendered into smart contract code. The smart contracts are highly modular, in that they are divided into components to perform specific functions, and thus can be individually managed. The smart contract code ultimately deployed to run on the blockchain is written in Node.js and Javascript [20].

Transactions created for we.trade utilize smart contract data sets (that is, templates), which reflect the rules and criteria pre-agreed between the parties. Once the smart contract data set is agreed between the buyer and the seller, and their banks (if financing products have been requested), the smart contract is generated on the we.trade platform. The smart contract will then scan for notification of the relevant settlement conditions. When such an event is detected, this will trigger the appropriate transaction processing, ensuring that the buyer and seller and their banks perform their respective responsibilities. Thus, the smart contract can automate the processing of the contractual clauses of traditional written contracts.

For example, an interface from a third-party transporter to confirm goods movement can trigger the delivery settlement condition, which can then launch the appropriate payment process. Other triggers which may be used to trigger payment processes include the initial agreement of the trade terms, the invoice being sent by the seller, the invoice being accepted by the buyer, the goods shipment confirmed by seller, and goods delivery confirmed by buyer. It should be noted that the buyer and seller, and their banks, will agree together when constructing the transaction whether the payment is to be fully automated, or whether the bank will still control the payment process.

The representative of a we.trade member bank mentioned in an interview that their buyers and sellers were at first sometimes 'scared' of the power of smart contracts. However, since these can be modified—with the approval of all parties—on the we.trade platform at any time before the relevant settlement condition is reached, the buyers and sellers subsequently became receptive to the use of smart contracts.

A we.trade smart contract data set will contain the following mandatory information:

- Buyer and seller details
- Purchase order reference
- Currency
- Amount
- Total price tolerance
- Buyer's bank and the buyer's bank account
- Seller's bank and the seller's bank account
- Delivery terms
- Payment terms
- Settlement conditions
- Expiry date

Where a bank financing product is selected, it will also contain this information:

- the goods or services involved
- the country of origin of any goods involved
- the price information (unit price, tax rate and quantity) of the goods or services

Smart contracts are indeed the core component of the we.trade platform, facilitating the automation of transaction processing and enforcement, and making possible the enhancements to competitive advantage that buyers, sellers, and banks can derive from participation in the we.trade platform.

3.5 we.trade Enabled Competitive Advantage

What are these prospective enhancements to competitive advantage for members?

According to interviews and media articles, feedback from buyers and sellers on the use of we.trade has been positive, with the value derived from membership in the platform found to be greater than expected. This value primarily results from the ability of we.trade to facilitate improved liquidity while reducing risk and enabling gains in efficiency, transparency, and security. These benefits as reported by buyers and sellers and their banks are further described in Table 4 [19,26–30,34–36].

Table 4. we.trade benefits.

Type	**Benefit**	**Description**
Greater Liquidity & Lower Risk	When there is no Letter of Credit	When it is not feasible to utilize a letter of credit, the we.trade platform can enhance the security and effectiveness of open account trade between buyer and seller. A letter of credit may be too expensive for an SME to afford, and may be too slow to process, for example when transport is made via ground within Europe.
	Enhanced Cash Flow	Buyers and sellers have enhanced access to the trade finance instruments offered by the banks on we.trade which can address cash flow needs. For example, the seller can use BPU Financing to receive payment earlier.
	When there is no Credit Insurance	When the limit for credit insurance has been reached, the added security provided by the we.trade platform may convince insurers to raise their limits, while at the same time the reduced amount of open receivables can reduce the amount of insurance required.
	Less Risk of Late Payment	Risk and impact to the seller of late payment is eliminated if the buyer arranges a BPU.
	Less Need for Prepayment	The seller may relax requirements for prepayment of contract amounts if the payment is guaranteed by the buyer's bank as a BPU.
	Extended Terms	The seller may grant the buyer other favorable terms, such as a longer payment period, if the payment is guaranteed by the buyer's bank.
	Automatic Settlement	Agreed settlement conditions between buyer and seller are automatically processed by the appropriate smart contracts upon encountering the relevant event triggers.
Efficiency	Streamlined Partnership Process	Buyers and sellers can safely establish their relationship digitally, with corroboration performed by their respective banks, while minimizing the need for onsite communications.
	Speed of Creation	The transaction can be created quickly between buyer and seller, and since any requests for financial products are now digital, they can also be processed by the respective bank much faster than was previously possible.
	Speed of Processing	Feedback on process events (e.g. a problem in delivery) is fast and online, with over 400 logistics providers on the we.trade platform.
	Paperless	The digital platform eliminates manual interventions, reducing the time and cost of processes, and removing opportunities for fraud.

Table 4 contd. ...

...Table 4 contd.

Type	Benefit	Description
Transparency	Manageability	Online visibility of process flows enables more effective management of the supply chain.
	Trust	Buyers and sellers can trust the we.trade platform and its mechanisms, whilst previously they may not have been able to trust each other.
	Reputation	Buyers and sellers have reported reputational benefits of credit worthiness and reliability arising from their membership on the we.trade platform.
	Immutability	Immutable supply chain transaction data would simplify the resolution of any disputes that could arise.
Security	Network	As a permissioned, private blockchain, the we.trade platform is secure against unauthorized access and manipulation.
	Identity	The bona fides of buyers and sellers are assured by the KYC processes performed by their banks during the onboarding process.
	Transaction	Verifiability of supply chain transaction data provides protection against fictitious orders, invoices, and payment requests.

Benefits for the participating banks correspond to the benefits to the buyers and sellers. The banks generate additional revenue streams with these new financing products for open account trade, and enjoy faster trade cycles, lower transaction costs, and lower risk of fraud due to efficient processing on the digital platform, with manual processes, paper documents, and the need for physical meetings minimized or eliminated.

The great majority of transactions (95%) on we.trade utilize BPU [22], to the reported mutual benefit of all parties to the transaction, the buyer, the seller, and their banks.

3.6 The Impact of COVID-19

The COVID-19 pandemic seriously impacted world trade, both operationally via border closures and quarantines, and financially through potential debt defaults and difficulties in receiving credit and insurance. A vicious circle began in which ever worsening company results made it ever harder for companies to be granted that credit and insurance. Many buyers and sellers were forced to find new trading partners, a task made even more difficult and risky by the restrictions on travel. With business processes in flux, new dangers arose from fraud.

Interviews with we.trade banks, buyers, and sellers suggest that the digital transformation enabled by we.trade became even more valuable to its members because of its role in mitigating the effects of COVID-19. The we.trade platform ensures that members have been properly vetted by their bank, the BPU instrument can mitigate the payment risk to sellers associated especially with new buyers, BPU Financing can overcome constraints on liquidity for sellers, and these factors can encourage sellers to be flexible in the terms they offer to buyers.

With traditional channels of communication and processes disrupted, the benefits of digital operations rapidly became apparent. David McLoughlin, Head of Commercialisation at we.trade in the first year of the COVID-19 epidemic, said that COVID-19, rather than inhibiting the move towards digital transformation, was in fact acting as a "catalyst" for it: "with COVID-19, the digitisation of trade and trade finance is now no longer a luxury, but a must" [37]. This finding was confirmed during interviews with we.trade banks, buyers, and sellers, in which it was suggested that even conservative companies traditionally resistant to change were becoming willing to move to a digital platform.

Meanwhile, SMEs in certain countries and industries have been receiving various forms of government financial support. It was mentioned by we.trade member banks in interviews that as such support concludes, the need for products such as we.trade's BPU and BPU Financing would become still greater.

3.7 Challenges

As a digital platform, we.trade must ensure that its ecosystem includes at least enough banks, buyers, and sellers, and generates at least enough transactions, to make the network viable in terms of the level of usefulness and value it provides its members. Only with such viability can the other important aspects of network effectiveness—such as the number, frequency, and size of connections, engagements, and relationships with existing and/or new partners—be meaningful.

In May 2020, the general manager of the we.trade JV gave a target of 25,000 users and 2.5 million transactions within three to four years [30]. With the volumes at the end of 2021 reported earlier in this chapter representing but a small fraction of these figures, it is not yet certain that we.trade can accumulate the needed scale. Without this, buyers and sellers may gravitate towards other digital solutions, and the banks may invest in those other business priorities which are in competition for management attention and resources.

The typical barriers to diffusion of blockchain solutions such as lack of knowledge and expertise [38] appear to be less of an obstacle in the case of we.trade, where the solution is promoted by the banks of the target users, that is, the buyers and sellers. So what challenges are constraining the growth of the we.trade platform?

As a multi-sided platform in which all participants must be members, we.trade is more susceptible to the network effect than are other platforms which do not require this (such as the Marco Polo platform). It can be said that we.trade has been suffering from the 'the-chicken-and-the-egg' dilemma:- buyers want sellers to be in place, and sellers want buyers to be in place, before either will agree to join the platform.

The country in which we.trade has the highest level of coverage of the trade sector is the Czech Republic. Whilst other countries have only one or two member banks each on the we.trade platform (and with these accounting for a minority of total trade in each country), the Czech Republic has three banks (ČSOB, Komerční Banka and Česká Spořitelna) which together serve approximately 80% of Czech companies engaging in trade [26]. This high level of coverage allows we.trade in the Czech Republic to perform even domestic trades. It will be instructive to trace the future development of we.trade in the Czech Republic in order to determine to what extent the number of banks and the trade coverage should be expanded in all markets. It should be noted that the banks in different markets will have varying reactions to the addition of national competitors to the we.trade platform.

While the onboarding process for new buyers and sellers to join we.trade is relatively fast, especially for the banks' existing customers, the onboarding process for new member banks is resource-intensive, involving numerous internal and external processes. Banks are by nature relatively conservative and understandably careful with their processes and must assign resources between many competing projects. These factors—the costs of affiliation as described by Loux et al. [39]—may represent a deterrent to new banks joining the we.trade platform.

In interviews, bank executives have acknowledged that adoption time is especially slow in the case of disruptive technologies. One bank representative stated that to convince the relationship managers in his own bank (who would propose we.trade to the bank's customers) of the value of the platform was a major challenge in itself. The significant effort required to on-board new banks onto we.trade (said in interviews conducted with bank executives to take at least one year, and up two years in the case of Akbank of Turkey) may then represent a significant constraining factor for the growth of we.trade.

A constraint affecting growth in the number of buyers and sellers may be the difficulty of expanding the scope of we.trade beyond Europe. While member buyers and sellers have expressed in the press and in interviews the desire to transact with partners in Asia and South America [40], we.trade has not yet published a timeline for the implementation of the complex process and governance changes needed to achieve this.

4. Conclusions

we.trade is an efficient, transparent, and secure digital platform that transforms the trade finance cycle for open account trade. The we.trade platform is enabled by a blockchain-driven governance structure of administrative, infrastructure, and application level mechanisms so as to coordinate and safeguard transactions which have traditionally been regulated manually via written contracts and social structures.

When the buyer and seller define their transaction on we.trade, they include the data set needed for generation of the relevant smart contracts, such as the goods/services description, pricing, delivery terms, payment terms, and the settlement conditions. The smart contracts then automatically process and enforce the parameters of the trade upon encountering the specified event triggers. Thus, the buyers, sellers, and their banks now place system-level trust [41] in the we.trade platform, replacing the need for trust in specific contracts or individuals.

This is a practical example of recent theoretical views of blockchain as a "new institutional technology that makes possible new types of contracts and organizations" [42] by possessing in itself the capability for the autonomous execution and enforcement of agreements between parties [43].

With the we.trade platform enabled by blockchain, its members can optimize their inter-organizational processes. This allows participating banks to create opportunities for enhancing competitive advantage via new and efficient revenue-generating services for the open account trade, while buyers and sellers can enhance their prospective competitive advantages because of their improved liquidity, reduced levels of risk, and greater efficiency in their trade processes. The digital transformation delivered by the we.trade platform has also helped to alleviate several of the serious barriers posed to trade by the COVID-19 epidemic. The major challenge now facing we.trade is to ensure that it reaches a viable scale of membership and transactions.

References

[1] International Chamber of Commerce. *2020 ICC Global Survey on Trade Finance*. International Chamber of Commerce, 2020.

[2] McKinsey. 2021. International Chamber of Commerce, and Fung Business Intelligence. *Reconceiving the Global Trade Finance Ecosystem*.

[3] Menichini, A. M. C. 2011. Inter-Firm Trade Finance in Times of Crisis. *The World Economy*, 34(10): 1788–1808. https://doi.org/10.1111/j.1467-9701.2011.01390.x.

[4] Deckert, A. 2021. Market Is Tight for Trade Credit Insurance; Policy Prices Inflated from Pandemic's Uncertainty. *Crisis Management Update*. https://crisismanagementupdate.com/market-is-tight-for-trade-credit-insurance-policy-prices-inflated-from-pandemics-uncertainty/. Accessed Dec. 7.

[5] Dhar, V. and Stein, R. M. 2017. FinTech Platforms and Strategy.
[6] Glaser, F. 2017. Pervasive Decentralisation of Digital Infrastructures: A Framework for Blockchain Enabled System and Use Case Analysis. *Proceedings of the 50th Hawaii International Conference on System Sciences.*
[7] Rochet, J.-C. and Tirole, J. 2004. *Two-Sided Markets: An Overview*. Institut d'Economie Industrielle, Paris, p. 44.
[8] Katz, M. L. and Shapiro, C. 1994. Systems competition and network effects. *Journal of Economic Perspectives*, 8(2): 93–115. https://doi.org/10.1257/jep.8.2.93.
[9] Gawer, A. 2014. Bridging differing perspectives on technological platforms: toward an integrative framework. *Research Policy*, 43(7): 1239–1249. https://doi.org/10.1016/j.respol.2014.03.006.
[10] Constantinides, P., Henfridsson, O. and Parker, G. G. 2018. Introduction—Platforms and infrastructures in the digital age. *Information Systems Research*, 29(2): 381–400. https://doi.org/10.1287/isre.2018.0794.
[11] Dhanaraj, C. and Parkhe, A. 2006. Orchestrating innovation networks. *Academy of Management Review*, 31(3): 659–669. https://doi.org/10.5465/amr.2006.21318923.
[12] Dyer, J. H. and Singh, H. 1998. The relational view: cooperative strategy and sources of interorganizational competitve advantage. *Academy of Management Review*, p. 21.
[13] Porter, M. E. 1980. *Competitive Strategy*. Free Press, New York.
[14] Rumelt, R. P. 1984. Towards a strategic theory of the firm. In *Competitive Strategic Management, Richard B. Lamb, ed.*
[15] Wernerfelt, B. 1984. A resource-based view of the firm. *Strategic Management Journal*, 5(2), https://doi.org/10.1002/smj.4250050207.
[16] Petersen, D. 2022. Automating governance: blockchain delivered governance for business networks. *Industrial Marketing Management*, 102: 177–189. https://doi.org/10.1016/j.indmarman.2022.01.017.
[17] Benoot, H. 2021. Interview with Hubert Benoot, Ex-Head of Trade Finance at KBC, and Ex-Chairman of the Board of Directors, We.Trade. Dec 20, 2021.
[18] Sangha, P. 2021. Interview with Parm Sangha, EMEA Blockchain Practice Leader, IBM. Oct 12, 2021.
[19] IBM. We.Trade Case Study. https://www.ibm.com/blogs/client-voices/we-trade-provides-intelligent-trading-solution/. Accessed Jul. 25, 2021.
[20] Gnagnarella, D. 2022. Trade Finance SIG Meeting - Hyperledger Foundation. *Hyperledger Foundation.* https://wiki.hyperledger.org/pages/viewpage.action?pageId=9110663. Accessed Feb. 24, 2022.
[21] Wragg, E. 2021. Blockchain Platform We.Trade Refills Its Coffers, CEO Departs. *Global Trade Review (GTR).* https://www.gtreview.com/news/fintech/blockchain-platform-we-trade-refills-its-coffers-ceo-departs/. Accessed Oct. 12, 2021.
[22] we.trade. We.Trade. https://we-trade.com/. Accessed Feb. 20, 2022.
[23] OECD. *Trade Finance for SMEs in the Digital Era*. Publication 24. 2021.
[24] we.trade. We.Trade User Manual / Application Guide v 2.9.
[25] we.trade. We.Trade Rulebook v 2.3.
[26] Morris, N. 2021. Automated Trade Payments Prove Popular for We.Trade Blockchain. *Ledger Insights - enterprise blockchain.* https://www.ledgerinsights.com/wetrade-blockchain-trade-finance-automated-payments/. Accessed Dec. 30, 2021.
[27] Hyperledger Foundation. Case Study: How We.Trade Helps Businesses Grow With Digital Smart Contracts Powered by Hyperledger Fabric. https://www.hyperledger.org/learn/publications/wetrade-case-study.
[28] Rao, R. 2019. First-Ever We.Trade Blockchain Platform Goes Live. techutzpah, Mar 15, 2019.

[29] Wragg, E. 2021. IBM Joins Banks as Shareholder in We.Trade. *Global Trade Review (GTR)*. https://www.gtreview.com/news/fintech/ibm-joins-banks-as-shareholder-in-we-trade/. Accessed Dec. 10, 2021.

[30] Ledger Insights. 2021. IBM Invests in Trade Finance Blockchain We.Trade as It Readies to Gear Up. *Ledger Insights*. https://www.ledgerinsights.com/ibm-invests-in-trade-finance-blockchain-we-trade-as-it-readies-to-gear-up/. Accessed Oct. 12, 2021.

[31] Blockchain Ireland. We.Trade Virtual Roundtable Discussion. 2021.

[32] we.trade. Leading The Way #1. *we.trade*. https://we-trade.com/resources/leading-the-way-1. Accessed Dec. 31, 2021.

[33] IBM. 2022. The Tradelens Platform. https://www.ibm.com/downloads/cas/AAREDOBM. Accessed Feb. 24, 2022.

[34] Perger, T. 2021. Interview with Tomas Perger, We.Trade Manager, Ceskoslovenska Obchodni Banka. Mar 12, 2021.

[35] Lopez, M. 2021. Interview with Michaël Lopez, Responsable Administratif et Financier, Ekoï. Dec 28, 2021.

[36] PYMNTS. We.Trade Executes First Corporate Transactions. https://www.pymnts.com/news/b2b-payments/2018/blockchain-trade-wetrade/. Accessed Dec. 1, 2021.

[37] Basquill, J. 2021. Blockchain Platform We.Trade Cuts Jobs Following Funding Slowdown. *Global Trade Review (GTR)*. https://www.gtreview.com/news/fintech/blockchain-platform-we-trade-cuts-jobs-following-funding-slowdown/. Accessed Oct. 12, 2021.

[38] Helliar, C. V., Crawford, L., Rocca, L., Teodori, C. and Veneziani, M. 2020. Permissionless and permissioned blockchain diffusion. *International Journal of Information Management*, 54: 102136. https://doi.org/10.1016/j.ijinfomgt.2020.102136.

[39] Loux, P., Aubry, M., Tran, S. and Baudoin, E. 2020. Multi-Sided Platforms in B2B Contexts: The Role of Affiliation Costs and Interdependencies in Adoption Decisions. *Industrial Marketing Management*, 84: 212–223. https://doi.org/10.1016/j.indmarman.2019.07.001.

[40] Nordea Bank. 2018. Transaction Banking Reflections with We.Trade.

[41] Hosmer, L. T. 1995. Trust: the connecting link between organizational theory and philosophical ethics. *Academy of Management Review*, p. 26.

[42] MacDonald, T. J., Allen, D. W. E. and Potts, J. 2016. Blockchains and the boundaries of self-organized economies- predictions for the future of banking. In *Banking Beyond Banks and Money*, Springer, Switzerland, p. 16.

[43] Lumineau, F., Wang, W. and Schilke, O. 2021. Blockchain governance — a new way of organizing collaborations?" *Organization Science*, 32(2): 47.

5

Investigating the Role of Permissioned Blockchains as Inter-Organizational Systems for Enabling Supply Chain Digital Transformation

Leo Yeung

1. Introduction

Blockchain was first introduced in Satoshi Nakamoto's white paper, "Bitcoin: A Peer-to-Peer Electronic Cash System" in 2008 [1]. While many people understood Blockchain as cryptocurrencies such as Bitcoin and Ethereum during the early stage of its development, the last few years have seen its adoption in business-to-business transactions. It slowly has become a digital technology that changes how business transactions are conducted between organizations and their trading partners with in trustless ecosystemst. However, some non-technology savvy executives do not fully understand blockchain technology. Hence, the adoption of Blockchain in business, including supply chain digital transformation, has been detrimentally impacted.

College of Business, City University of Hong Kong, 93 Tat Chee Avenue. Kowloon Tong, Hong Kong.
Email: leoylw@netvigator.com

Blockchain can be categorized as being Permissionless (e.g., Bitcoin) and Permissioned. Permissioned Blockchain including Federated/ Consortium and Private Blockchain, can only be accessed by pre-approved participants. The sharing of transaction data is controlled among the participants. Therefore, it is suitable for mission-critical business transactions including those for supply chain digital transformation. Permissioned Blockchains are the latest generation of Inter-organizational Systems (IOS). It is equipped with the characteristics of IOS but evolved into a decentralized model without the need for a trusted intermediator.

From an adoption perspective, it is important to understand the characteristics of Permissioned Blockchain, the business models which support its, as well as its industrial applications. Moreover, it is also crucial to understand what drives the adoption of Permissioned Blockchain technologies. This chapter aims to enhance understanding in this specific area.

This chapter introduces Permissioned Blockchain as one of the two categories of Blockchain. Unlike Permissionless Blockchains which operate via a decentralized business-to-consumer or consumer-to-consumer model, Permissioned Blockchains are a decentralized business-to-business model which enables supply chain digital transformation. The next section of this chapter defines what Permissioned Blockchains are; explains how the concept differs from Permissionless Blockchains; and why it is important to differentiate Permissionless and Permissioned Blockchains. IOS are information systems used by organizations and their business partners for exchanging business transaction data and orchestrating business processes with each other. IOS is important for supply chain digitization because a supply chain ecosystem consists of lots of trading partners transacting supply chain data with each other. Blockchain on the one hand enables these transactions. On the other hand, it allows the ledger of the data to be kept in a decentralized model without relying on a trusted intermediator. Therefore, Permissioned Blockchain is considered the latest generation of IOS. The third section of this chapter explains what IOS is; why Permissioned Blockchains are the latest generation of IOS; and how it is adopted for supply chain digital transformation. Permissioned Blockchain is viewed as the new generation of IOS and its success depends on an organizations' willingness and ability to adopt. It is therefore important to understand Permissioned Blockchain from an adoption perspective. Section 4 introduces two specific Permissioned Blockchain adoption models. The first one focuses on individual-firm adoption and the second one focuses on adoption through a Blockchain Industry Consortium model. Section 5 lists the determinants of Permissioned Blockchain adoption. Six factors are identified and categorized into the Technological, Organizational, Inter-organizational, and Environmental context referencing the TOE Framework. These factors were ascertained through interviews with 10 senior information technology (IT) executives in Hong Kong, which is one of the fasted growing blockchain markets in the world. Senior IT executives interviewed included both adopters and non-adopters of blockchain technology (Section 6).

2. Permissioned Blockchain

2.1 What is Permissioned Blockchain

Blockchain is defined as "a cryptographic, or encoded ledger—a database of transactions in the form of blocks arranged in a chain. These are validated by multiple users through consensus mechanism (such as proof-of-work in Bitcoin mining) shared across a public or private network."[1] It is a decentralized ledger with transaction data encrypted and validated through mutual consensus of participants. Blockchain is characterized by its nature of decentralization, persistency, anonymity, and audibility [2]. The advantage it offers to users in validating transactions without relying on any trusted intermediator does not exist in previous technologies. With this advantage, Permissioned Blockchain can be applied in many aspects of businesses. It is forecasted by the World Economic Forum that approximately 10% of global GDP will be stored in Blockchain technology by 2027.[2]

Blockchain can be categorized into Permissionless Blockchain and Permissioned Blockchain [3]. Permissionless Blockchain are also called Public Blockchain. It is open for the participation of anyone with the anonymity of his/her identity. Permissioned Blockchain, on the other hand, can only be accessed by pre-approved participants. The sharing of transaction data is controlled among the participants and is suitable for mission-critical business transactions [4]. It is divided into Federated/ Consortium and Private Blockchain by some researchers [5,6]. To enable the execution and completion of business transactions, Permissioned Blockchain makes use of distributed Ledger technology (DLT), Smart Contract, and Decentralized Applications (DApp) [7]. There is a wide range of business use cases based on Permissioned Blockchain across different industries. Examples include tracking items across the supply chain with multiple handlers for anti-counterfeiting and status monitoring, exchange of patient information in the healthcare system, exchange of information

---

[1] Source: https://www.everestgrp.com/2016-05-blockchain-technology-bfsi-benefits-market-insights-20805.html/.

[2] Source: http://www3.weforum.org/docs/WEF_GAC15_Technological_Tipping_Points_report_2015.pdf.

in financial products from application through approval to a settlement which is also categorized as Decentralized Finance (DeFi) [8].

2.2 Technical Differences between Permissionless Blockchain and Permissioned Blockchain

There are several key differences between Permissionless Blockchain (Public) and Permissioned Blockchain (Federated or Private) from technology's perspective. They can be observed in the below table.

As indicated in Table 1, there are 5 core differences between Permissionless and Permissioned Blockchains in terms of the condition of participation, access control, identification of participants, consensus mechanism, and the control of the source code.

From the perspective of participation, access control, and identity, Permissionless Blockchain is open for anyone to participate as a node or a "miner" to validate transactions and compete for reward in terms of a crypto coin in return. Anyone can read and write transaction data into the block after being validated. The participant's identity is pseudonymous and cannot be identified as it just comes from a dynamic IP address. On the other hand, participants of Permissioned Blockchain are by invitation only and must be pre-approved. They can be organizations with common interests or purposes. For example, they may all come from the same industry or the same business ecosystem. As such, the read and

Table 1. Comparison between Public Blockchain and Permissioned Blockchain

	Public	Permissioned
Participation	Anyone can participate	By invitation only, vetted either by a central authority, consensus, or other criteria
Access Control	Anyone can read and anyone can write (subject to validation)	Read and write access may be restricted to protect the privacy of data
Identity	Pseudonymous	Participants identified, preferably strongly
Consensus	Typically requires a majority of validator nodes	Maybe done by a smaller set of nodes, such as stakeholders and/or knowledge-holders
Control of Code	Anyone can make changes, but a majority of nodes decide which to keep	Mayabe centralized or controlled by a consortium

Source: McBeath B (2018), Blockchain, Identity, and CSR in 2018[3]

[3] Source: https://www.experfy.com/blog/blockchain-identity-and-csr-in-2018.

write authority of participants are also restricted by the operators of the network nodes which could be an individual organization or a group of organizations in the form of a consortium.

In addition, a major feature of Blockchain is the consensus mechanism. Consensus is the process by which participants on the Blockchain peer network agree on confirming transactions [9]. These are the rules and techniques for the participants to reach a consensus on the realness of transactions. For Permissionless Blockchain, the consensus is Proof-based such as the Proof-of-Work protocol. Participants join Permissionless Blockchain as Blockchain network node operators. They must solve a difficult cryptographic question to prove their trustworthiness which some people called "mining" before they can add new blocks to the chain. The consensus is based on the "right answer" that came up by over 50% of the participants despite the reward only going to the participant who took the shortest time to answer. Likewise, the majority of the participants can vote and decide on change or enhancement in the source code of the Blockchain. This governance of the source code can be achieved through a decentralized autonomous organization (DAO) which is voted by the participants [10].

As the participation of Permissioned Blockchain is by invitation only, it is a closed user community with all known participants. Validating nodes in Permissioned Blockchain do not need to prove their trustworthiness as they are pre-qualified. Therefore, these known participants engage in Fault-Tolerant Consensus to ensure the consistency of all replicated copies of the ledger under their management. Fault-Tolerant Consensus can be based on a variety of consensuses such as Proof-of-Authority, PBFT, or other distributed consensus protocols [9]. It is performed by the participants who are also the stakeholders of the Blockchain network. Likewise, the control of code changes is owned by the stakeholders.

2.3 Different Business Models of Permissionless and Permissioned Blockchain

Permissionless and Permissioned Blockchains have different technical characteristics as mentioned in Section 2.2. These differences boil down to the key question of what business models they are supporting. This question is important because a mismatch between the Blockchain technologies and the business models leads to a failure in the adoption of Blockchain.

Permissionless Blockchain is relevant for the business-to-consumer or the consumer-to-consumer model. It enables consumers to freely transact with each other including those with low trust levels without the use of a trusted intermediator to validate the ledger. The transaction of cryptocurrencies such as Bitcoin is a typical example. Conversely,

Permissioned Blockchain with pre-approved participants is relevant for transacting business mission-critical data in a business-to-business (B2B) model. This model transacts across a closed user community like IOS such as EDI and e-Supply Chain applications [11]. In addition, it enables the transactions processed in a decentralized model. Hence, it is the Permissioned Blockchain's decentralized business-to-business model which is relevant for supply chain digital transformation projects. The differences between Permissionless and Permissioned Blockchain call for the focus of supply chain digital transformation projects on the characteristics of Permissioned Blockchain. With this focus, there are two key questions that the adopters of Permissioned Blockchain should consider. The first question is what is the form of adoption? Form of adoption is an irrelevant topic for Permissionless Blockchain because it is an end-user level adoption. However, it is a crucial one for Permissioned Blockchain as it impacts the success and effectiveness of the adoption. There are two forms of adoption. They are individual-firm adoption and adoption through Blockchain Industry Consortium. These two forms of adoption will be discussed in Section 4 of this chapter.

The second question is what factors drive the adoption? While it is the end-user to decide on Permissionless Blockchain adoption, it is the organization to decide on adopting Permissioned Blockchain. Therefore, the adoption determinants should be those influencing corporate decisions rather than end users' decisions. These adoption determinants will be introduced in Section 5 of this chapter.

3. Permissioned Blockchain as Inter-Organizational Systems

3.1 What is Inter-organizational Systems

As mentioned earlier, IOS refer to computer systems deployed by two or more organizations together to exchange information across the organizational boundaries [12,13]. Organizations use IOS to share information, exchange data, and orchestrate business processes for collaborating with trading partners on mission-critical business tasks [14]. IOS have strong relevance for supply chain digital transformation as they are tools to integrate the systems of supply chain participants such as buyers, sellers, as well as logistics, and financial service providers for executing and monitoring the supply chain. [14].

The deployment of IOS can be bi-lateral in the form of "peer-to-peer" or multi-lateral in the form of "one-to-many" or "many-to-many." However, the most effective implementation strategy is "one-to-many", or called the "hub-and-spoke" model, which is for one "hub" which has strong negotiation power over the "spokes" it transacts with to implement the corresponding IOS. Before the presence of Permissioned Blockchain, Electronic Data Interchange systems (EDI) or Business-to-Business applications building on top of EDI transactions, are the most popular IOS for supply chain integration [15,16]. For instance, a buyer implements an EDI-based system and requests its suppliers to implement the corresponding EDI-based system so that they can transact supply chain information digitally and automatically. This model is popular in the retail industry with big retailers such as Walmart.[4] Vice versa, a strategic supplier can also request its buyers to do the same. This can be seen in the semi-conductor industry with Intel as an example. The semi-conductor company has been promoting the use of RosettaNet, an XML-based technology, for supply chain integration.[5]"Peer-to-peer" model can be observed in the logistics industry with carriers, 3PL, and freight forwarders to exchange logistics-related information peer-to-peer or point-to-point.

Quality, correctness, completeness, and reliability are crucial data integrity in keeping the information shared by the organizations in the same IOS environment [17]. Such data integrity is ensured by a trusted intermediator which can be the "hub" in the "hub-and-spoke" model. For instance, a buyer who sends purchase orders to its suppliers is the "hub." It will keep track of the orders and order fulfillment status. In the case of a "peer-to-peer" model, it will base on the reconciliation between the sender and the receiver, i.e., the buyer and the supplier. If the transactions are outsourced to a third-party information service provider, then the data integrity will be ensured by the service provider. An example of a third-party information service provider is Opentext Business Network[6] which provides a cloud platform for IOS outsourcing. A diagrammatic illustration of the two models is depicted in the figure below.

The advantages of adopting IOS are different from those of adopting internal IT systems. Adopting IOS can not only improve the performance of an organization but also improve the collective performance of all organizations participating [13]. Adopting IOS is also a competitive strategy by extending an organization's scope of business into the business of its business partners [12]. By adopting IOS, an organization can compete with its entire business ecosystem involving external trading partners.

The importance of IOS is increasing with the increasing disruptions in many organizations' supply chains. These disruptions could be caused

[4] Source: https://walmartsupplychain.weebly.com/edi-system.html.

[5] Source: https://www.intel.com/content/www/us/en/supplier/resources/technologies/simulation-demo-condensed-eng.html.

[6] Source: https://businessnetwork.opentext.com/.

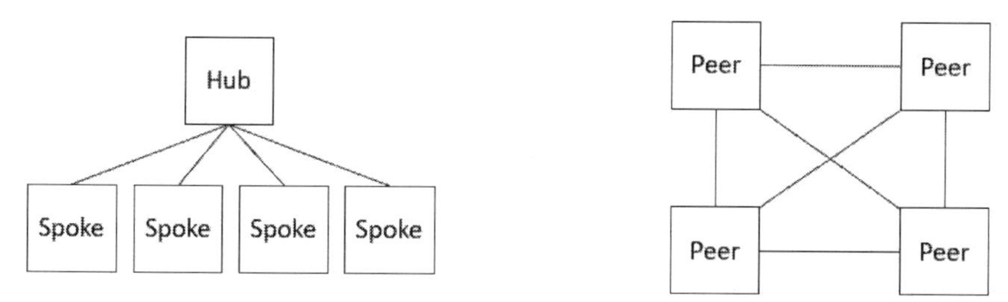

Figure 1. Hub-and-spoke Business Model & Peer-to-Peer Business Model.

by competition, extreme weather, trade restrictions imposed by countries, the latest COVID-19 pandemic, etc. Organizations are advised to diversify the trading partner base in their supply chain to mitigate the risk of being disrupted [18]. For example, to broaden the supplier base by recruiting suppliers from different countries instead of just from a few locations. However, the more diverse one's supply chain ecosystem, the higher the dependence of IOS to ensure the resilience of the supply chain.

3.2 Permissioned Blockchain as the latest generation of IOS

As explained in Section 3.1, IOS are the systems enabling transactions across organizations. Permissioned Blockchain enables transactions to be kept in blocks and shared between pre-approved participants. Therefore, it processes the characteristic of IOS in terms of facilitating the creation, storage, transformation, and transmission of information [13]. In addition, IOS are used for orchestrating business processes across trading partners. Implementing Permissioned Blockchain is the deployment of the technology for digitizing business processes whether it is exploitative or exploratory [19]. Hence, it also serves the purpose of IOS in terms of process orchestration. However, there is a key feature of Blockchain not found in previous IOS which is the distributed trust without relying on a trusted intermediator to validate transactions. Transaction data are recorded in the blocks after being validated by the corresponding Blockchain network node operators in a decentralized model. Therefore, Permissioned Blockchain can be perceived as the latest IOS. It improves the collaboration between trading partners as well as the efficiency and visibility of transactions between them [20].

There are two adoption approaches to deploy Permissioned Blockchain as IOS. They support the "hub-and-spoke" model and the "peer-to-peer" model in IOS introduced in Section 3.1.

For the "hub-and-spoke" model, the organization is to operate all the Permissioned Blockchain network nodes for centralizing the decentralized. Some researchers called this a Private Blockchain. The second model is to implement through a Blockchain Industry Consortium as "peer-to-peer." It is also called Consortium Blockchain. These two models will be explained in more detail in Section 4.

3.3 Permissioned Blockchain as IOS for Supply Chain Digital Transformation

An end-to-end supply chain involves activities in planning, making, buying, moving, and selling of products across different stakeholders. Each of them contributes transaction data to the chain. In return, they consume the transaction data for their business purposes. Stakeholders in the supply chain can be described in three groups:

1. *Buyer and Seller:* to buy/sell products. The exchange of title of the product is based on evidence of the product's authenticity, as well as of the fulfillment of trading conditions. This evidence includes information transactions such as purchase orders, advanced ship notices, receiving advice, invoice, remittance advice, etc.
2. *Logistics Services Providers:* to move the product from the origin to the destination. They include warehousing operators, truckers, freight forwarders, sea or air carriers, and customs service providers. Each of them contributes and consumes shipment status transaction data while delivering supply chain logistics services.
3. *Financial Services Providers:* to provide financial services for facilitating trade transactions. They are banks or factoring service providers.

Traditionally, information created in the activities of the three groups of stakeholders is provided either in unstructured formats (e.g., paper, email, email attachment, fax, etc.) or structured formats (e.g., EDI, XML, etc.). It is possible to validate bilaterally transaction data either manually or digitally by using IOS. The validation will be less easy if the transaction data involves, or passes through, more than 2 parties. An outsourcing service provider can provide syntax-level data validation, but it is the data creator's database to provide the true original data.

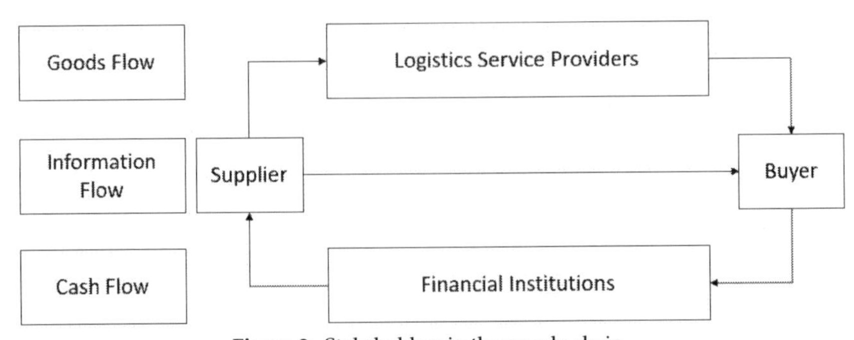

Figure 2. Stakeholders in the supply chain.

Adoption of Permissioned Blockchain as IOS for supply chain digitization brings value to all three groups in ensuring the accuracy and trueness of the transaction data.

For *buyer and seller*, the ultimate objective is to enable the transfer of the product's title by executing what has been agreed and stated in the Smart Contract. By doing so, intermediates and their effort to perform the collection, validation, and reconciliation of the trade information can be eliminated. Transfer of title is executed once the transaction data stored in the Permissioned Blockchain fulfills the trade conditions stated in the Smart Contract. It will take time to make this automation happen as significant system and business process changes are required. Even before that, it is useful for the inclusion of production-related data to be stored in Permissioned Blockchain to ensure the product authenticity and quality of the product, as well as the condition in manufacturing or assembling, are accurately & completely recorded. This is especially important for luxury goods or food products. The latest example is the new requirement of proof of the origin by the US Customs for importing cotton and cotton fashion items into the US.[7] It is also crucial to ensure counterfeiting can be prevented for life-critical products such as PPE used during COVID-19 pandemic [21]. Permissioned Blockchain can be used for tracking the origin of the shipment. In addition, the adoption of Permissioned Blockchain may include the adoption of the Internet of Things (IoT) to create the digital twin for the product. This can provide dynamic monitoring of products' condition during in-transit.

For *Transport & Customs related stakeholders*, Permissioned Blockchain can capture both the location of the product and the condition of the product while it is in transit from the original to the destination. This can be achieved by either the logistics provider submitting that information statically or to incorporate the IoT technology, either at the vehicle level or at the shipment level, to communicate directly with the Permissioned Blockchain platform for dynamic logistics tracking. For example, a Swiss manufacturer of pharmaceutical containers called SkyCell provides containers with IoT sensors to track the temperature and concussion data of the container and send those data to their Permissioned Blockchain platform. SkyCell's containers have been used to transport COVID-19 vaccines.[8]

---

[7] Source: https://www.mayerbrown.com/en/perspectives-events/publications/2021/01/us-bans-imports-of-all-products-containing-cotton-and-tomatoes-from-chinas-xinjiang-region.

[8] Source: https://www.skycell.ch/news/skycell-increases-its-availability-for-in-the-us-market-with-a-new-partnership/.

Permissioned Blockchain was also used for managing financial transactions with freight carriers. For instance, Walmart Canada used Permissioned Blockchain for managing invoices from and payments to 70 3rd party freight carriers which transport over 500,000 shipments for them per annum.[9] Those freight invoices included comprehensive shipment & tracking data which require massive manual reconciliation before being processed in Permissioned Blockchain.

For *Financial institutions* that require the applicants to provide their trade transaction records for processing the trade financing applications, a lot of validation work is needed. For instance, the Invoice has to be reconciled with its corresponding Purchase Order and Receiving Advice. Before Permissioned Blockchain, applicants spent a lot of effort to make this information available to the trade financial institution. With Permissioned Blockchain, Orders related transaction data is stored in the Blockchain which the data's immutability is ensured. The financial institution can consume data in the Blockchain in their trading financing application. It speeds up the application process and saves the cost of processing. Standard Chartered Bank, as an example, partnered with Trusple, in integrating its Straight2Bank digital banking platform with the latter's Permissioned Blockchain platform for trading financing.[10] The bank has also joined a Blockchain Industry Consortium called Contour for completing Letter of Credit transactions over Permissioned Blockchain.

4. Adoption Models of Permissioned Blockchain

Permissioned Blockchains bring additional value to business-to-business processes including supply chain digitization over the previous IOS. It is therefore essential to understand its adoption. For instance, how it is adopted and what drives adoption. There are two models to adopt Permissioned Blockchain as mentioned in Section 3.2. The first model is for an organization to adopt it individually. The second one is to adopt it collaboratively with other industry players in the form of an industry consortium. To understand corporate's considerations in Permissioned Blockchain adoption, a round of executive interviews with senior IT executives was conducted in Hong Kong to draw their insights. The interviews confirmed the existence of both forms of adoption. A discussion on the insights is provided in Section 6 of this chapter.

---

[9] Source: https://hbr.org/2022/01/how-walmart-canada-uses-blockchain-to-solve-supply-chain-challenges?utm_medium=email&utm_source=newsletter_daily&utm_campaign=dailyalert_notactsubs&deliveryName=DM169499.

[10] Source: https://www.sc.com/en/media/press-release/weve-completed-the-first-live-transaction-on-new-blockchain-enabled-trading-platform-trusple/.

4.1 Individual-firm Adoption of Permissioned Blockchain

Individual-firm adoption of Permissioned blockchain simplest model of adoption [22]. The way it works is for an organization to manage the ledger and operate all the Blockchain network nodes by itself. The organization is the trusted party of the ledger. One may argue that if individual-firm adoption model complies with the decentralized principle of Blockchain because the technology is decentralized but the control of the ledger is not [7]. Nevertheless, other characteristics of Permissioned Blockchain in terms of immutability and integrity of data in the record of transactions are what individual adopters see of high value to their business [9].

Individual-firm adoption of Permissioned Blockchain as IOS is possible as long as the implementing organization has the cooperation and in many cases influence or control, of its trading partners. For instance, it may be a strategic buyer, supplier, or service provider to its trading partners. In this case, this is the "hub-and-spoke" model mentioned in Section 3.2.

This model is relevant for supply chain digital transformation in two areas. The first one is in the order management process between buyer and seller. Permissioned Blockchain can be used by buyers or sellers to ensure product authenticity, to keep track of order fulfillment status, and eventually automate the change of product's title by executing the smart contract. The second one is for the organization to facilitate the transactions across its trading partners with information supplemented by the logistics and financing service providers. For example, the organization may work with a warehouse operator and a logistics service provider. Goods are to be shipped in/out of the warehouse by the logistics service provider. Hence, the organization as a shipper must have accurate shipment status from the warehouse and the logistics service provider for inventory management. Permissioned Blockchain can be adopted for this purpose. The limitation of individual-firm adoption is the inability to extend the verification of transaction data beyond the direct trading partners in the blockchain network.

4.2 Adoption through joining Blockchain Industry Consortium

4.2.1 What is a Blockchain Industry Consortium

A Blockchain Industry Consortium is a business model for industry players, particularly the market leaders in the industry, to get together for adopting an innovation collaboratively. An example is the consortium-based "industry exchanges" formulated in 2000–2001 to facilitate supply chain integration across trading partners [23]. There were at least eighteen of them at one time in industries such as transportation, retail, automobile, electronics, healthcare, and others. Some are still existed today such as the

Global Healthcare Exchange, e2Open, Covisint (acquired by Opentext), Elemica, etc. Another type of consortium is to promote the adoption of technology standards across the industry. An example is RosettaNet which included a group of semi-conductor and hi-tech manufacturing companies to develop and promote a business-to-business standard called RosettaNet Implementation Framework (RNIF) aiming at replacing the use of Electronic Data Interchange (EDI) in their supply chain ecosystem [24].

A key objective of forming a Blockchain Industry Consortium is for the participants to adopt Permissioned Blockchain collectively for minimizing costs & risks as well as reaching the critical mass in the industry quickly [25]. They may subsequently work on the standards of Blockchain Network integration for interoperability. Moreover, they have an additional role which is to share the ledger and control the governance of the Blockchain network in terms of the consensus mechanism and code changes. Furlonger and Kandaswamy (2019) defined Blockchain Consortium as

"A blockchain consortium is a group of companies joining forces to foster cooperation by sharing a ledger and by defining standards and a governance model to create a digital ecosystem to reduce operational risk, minimize costs or enhance customer service" [26].

There are different categories of Blockchain Consortium which can broadly be divided into Technology-centric or Business/Industry-centric. Technology-centric Consortium has an interest in adopting the technology, sharing knowledge and best practices, promoting technology standards, as well as influencing the future development of the technology. For example, Hyperledger, Enterprise Ethereum Alliance, and R3 are examples of Technology-centric Consortiums.

Business/Industry-centric Blockchain Consortium, or what we called Blockchain Industry Consortium, is formed by the industry players for adopting Permissioned Blockchain collaboratively. The objective of forming Blockchain Industry Consortium is to solve problems that impact the industry and can only be addressed at the industry's level. Participants in the Blockchain Industry Consortium are from the industry ecosystem horizontally and vertically. Horizontal participants are providing similar products or services in the same industry, so literally, they are competitors. However, they are co-operating in the consortium for mutual benefits. This co-existence of both competition and co-operation inter-organizational relationship is defined as "co-opetition" [27]. Such "co-opetition" is unique in the Blockchain Industry Consortium. Whether the industry players can strike a good balance between co-operation and competition in the consortium has a profound impact on the success of the consortium.

Table 2. Examples of permissioned blockchain industry consortium.

Consortium	Area of interest	Participants
TradeLens https://www.tradelens.com/	Shipping solutions	8 shipping lines including Maersk, CMA CGM, MSC, Hamburg Sud, PIL, Zim Lines, ONE, Hapag-Lloyd as well as container terminals and freight forwarders
Global Shipping Business Network https://www.cargosmart.ai/en/solutions/global-shipping-business-network/	Shipping solutions	5 shipping lines including CMA CGM, COSCO, Evergreen Marine, OOCL, and Yang Ming Lines as well as 4 terminal operators including DP World, Hutchison Ports, PSA International, and Shanghai International Port.
we.trade https://we-trade.com/	Trade Finance	13 banking partners including CaixaBank, UBS, Nordea, Erste Group, Rabobank, HSBC, KBC, UniCredit, Natixis, Societe Generale, Santander, Eurobank, Deutsche Bank,
MarcoPolo https://www.marcopolo.finance/	Trade Finance	17 banks including Raiffeisen Bank International joins BNP Paribas, Commerzbank, ING, LBBW, Anglo-Gulf Trade Bank, Standard Chartered Bank, Natixis, Bangkok Bank, SMBC, Danske Bank, NatWest, DNB, OP Financial Group, Alfa-Bank BayernLB, Helaba; S-Servicepartner and Raiffeisen Bank International
Voltron https://www.voltrontrade.com/	Trade Finance	8 founding banks including SEB, CTBC Bank, HSBC, BNP Paribas, NatWest, Bangkok Bank, Standard Chartered Bank, ING, and over 40 participant banks.
eTradeConnect https://www.etradeconnect.net/Portal	Trade Finance	7 initiating banks including ANZ, BOC, BEA, Hang Seng Bank, HSBC, Standard Chartered Bank, DBS Bank, and also 5 other banks including Agricultural Bank, Bank of Communications, BNP Paribas, Shanghai Commercial Bank, and ICBC.
Contour https://www.contour.network/	Trade Finance	11 banks including Bangkok Bank, DBS, Citi, HSBC, ING, SEB, BNP Paribas, CTBC Holding, HDBank, SABB, Standard Chartered.
MOBI https://dlt.mobi/	Automotive	Auto companies include BMW, Bosch, Ford, General Motors, Groupe Renault, ZF, Aioi Nissay Dowa Insurance Services USA

The Blockchain Industry Consortium model is a popular way for adopting Permissioned Blockchain in its introductory stage. There was only 1 Blockchain Industry Consortium in 2014 but there were 53 active consortia in 2021.[11] The below table shows examples of the more representative ones.

4.2.2 Adopting through Blockchain Industry Consortium

The adoption of Permissioned Blockchain through the Industry Consortium enables an organization to digitize its inter-organizational business processes through participating in one or more Blockchain Industry Consortium. Distributed Trust, in addition to the immutability and integrity of data in the record of transactions, is crucial in this model. It is because these participants, despite being pre-approved and jointly executing the consortium governance, are competitors with each other in the same industry, and therefore the level of trust is low.

The implementation of Permissioned Blockchain in the form of Industry Consortium adopts the "peer-to-peer" model mentioned in Section 3.2. An example of an Industry Consortium model is referred to as a Contour. This consortium consists of 9 global banks aiming to digitize the Letter of Credit process using Permissioned Blockchain.[12] In international trade, a bank can in one instance issue Letters of Credit but in an other instance act as an endorsing bank for other transactions. It provides data to another bank in one transaction and receives data in another transaction. Therefore, it makes good sense for a peer-to-peer business model in the Industry Consortium. Another example is TradeLens which was set up by Maersk and joined by 4 other leading ocean carriers as a Blockchain Industry Consortium targeting the ocean transportation-based supply chain.[13] The consortium serves a supply chain ecosystem consisting of ocean carriers, container terminals, inter-modal providers, container depots, and customs authorities. Likewise, a carrier in an ocean alliance provides data to another carrier in the same alliance on one shipment, it receives data on another shipment.

---

[11] Source: https://www.blockdata.tech/blog/general/the-state-of-blockchain-consortiums-in-2021.
[12] Source: https://contour.network/.
[13] Source: https://www.tradelens.com/.

5. Adoption Determinants of Permissioned Blockchain as IOS

5.1 Conceptual Model of Permissioned Blockchain Adoption

As Permissioned Blockchain is a crucial IOS for supply chain digital transformation, it is important to understand what factors determine the adoption of Permissioned Blockchain as mentioned in Section 2.3. It is useful to understand these determinants through a relevant conceptual model.

The Technological, Organizational, and Environmental (TOE) framework is used to assess innovation adoption at the organizational level [28]. It is a model that can be applied to understand the adoption determinants of technologies at the organizational level such as EDI, e-Business Applications, ERP systems, and e-Supply Chain Management Systems, etc. [11,29,30]. It is, therefore, appropriate to apply the TOE Framework as a conceptual model for understanding the adoption determinants of Permissioned Blockchain.

In the TOE Framework, all external factors are categorized under the Environmental context. This chapter suggests separating inter-organizational related determinants from the Environmental context. This is because inter-organizational factors are distinctive for Permissioned Blockchain as IOS. Inter-organizational factors involving parties have a direct relationship with the adopting organization such as the trading partners and the industry consortium. These parties are involved in the end-to-end inter-organizational business processes. Parties indirectly related to the organization such as competition and government policies should still be categorized as environmental factors. A literature review identified 65 possible determinants identified. Six of these most significant one identified will be discussed in Section 5.2. A conceptual model of Permissioned Blockchain adoption is developed and presented in the below figure.

5.2 Adoption Determinants of Permissioned Blockchain as IOS

As shown in the above figure, this conceptual model referenced the TOE Framework but with an inter-organizational dimension added. Six determinants categorized into the four contexts were identified for Permissioned Blockchain adoption and ascertained by 10 senior IT executives interviewed. They are discussed in the below section.

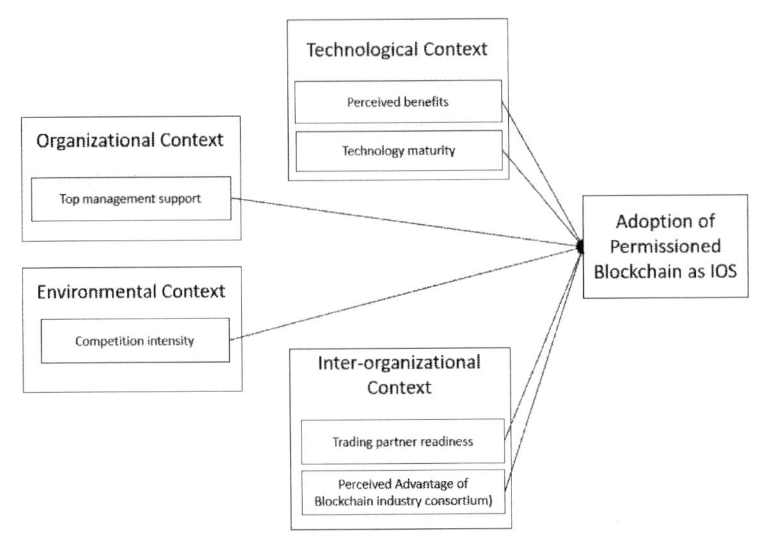

Figure 3. Conceptual model of permissioned blockchain adoption.

5.2.1 Technological Context

Perceived Benefits

Perceived Benefits are the advantages an organization expects to receive [31]. They include both direct benefits such as revenue improvement and cost reduction, as well as indirect benefits such as improvement in operational efficiency, customer relationship, competitiveness, and corporate image [11,32,33]. While actual benefits are important for adopting an innovation, Perceived Benefits are more relevant for Permissioned Blockchain adoption decision because the decision is based on the benefits expected whereas the actual benefits can only be obtained after the adoption. Perceived Benefits were considered as determinants in Blockchain adoption research based on TOE [34, 35].

Technology Maturity

Technology Maturity refers to the extent of the functional characteristics of a technology being proven by other adopters [36]. Lee et al. (2017) identified three parameters to evaluate Technology Maturity. They are if it has been adopted successfully by other adopters; the benefits expected are demonstrated, and the weaknesses or issues are known and addressed. The more successful adoptions of the technology, the higher is its maturity. Technology Maturity is important for Permissioned Blockchain adoption as the adoption is a corporate decision. Organizations always observe the adoption of other organizations as a reference for their adoption decision [37]. This can also be observed from the executives we interviewed.

5.2.2 Organizational Context

Top Management Support

Top Management Support refers to the commitments of the top management in adopting the innovation. The top management has to demonstrate its commitment by including the innovation in its corporate strategy, also investing resources, taking risks, as well as acting as the role model of adoption [38]. For the adoption of Permissioned Blockchain as IOS, top management support also includes the top management's drive to incorporate it into the inter-organizational business processes [30]. Top management Support is crucial for Permissioned Blockchain adoption, especially in its initial phase. It is because the benefits and risks are unclear during this stage so it would be challenging to justify adoption purely based on the estimation of costs & benefits. It takes the top management to make the adoption decision in front of the uncertainties. Top management support is also crucial in determining the adoption model.

5.2.3 Inter-organizational Context

Trading Partner Readiness

Trading Partners refer to the suppliers, buyers, and logistics or financial service providers in the supply chain of the organization that intends to adopt Permissioned Blockchain [16]. Trading Partner Readiness is important because the success of IOS adoption relies on if the organizations and their trading partners are well prepared [31]. The IOS and the related inter-organizational business processes cannot be executed if the trading partners are not ready. Hence, it is a boundary condition for adoption [39].

There are two aspects of trading partner readiness including both technical and business [40]. Technical readiness is if a trading partner's IT systems are prepared to integrate and transact with the adopter's Permissioned Blockchain platform. Business readiness is if a trading partner is prepared to make changes in its business processes aligning with the processes of the adopter after Permissioned Blockchain is adopted.

Perceived Advantages of the Blockchain Industry Consortium

As explained in Section 4.2.1, Blockchain Industry Consortium is a collaborative adoption model of Permissioned Blockchain including organizations in the same industry [26]. Blockchain Industry Consortium is relevant for Permissioned Blockchain adoption because the pre-approval participants in the consortium are required to share the ledger and to operate the Blockchain network's governance model on consensus mechanism and code changes [26]. In addition, there are perceived advantages of Blockchain Industry Consortium which drive and expedite adoption. Those perceived advantages include minimizing the adoption cost and risk, improving operation efficiency, and accererating the speed to get to the critical mass. These cannot be achieved by individual-firm adoption [23].

5.2.4 Environmental Context

Competition Intensity

Competition Intensity is the severity of competition in the industry in which the organization operates [41]. Adopting innovation has always been a strategy that organizations use to improve market competitiveness [42]. From a Permissioned Blockchain perspective, it supports new business models and enables business processes can be digitized. These improve an organization's efficiency and cost-effectiveness which are important to compete in a competitive market. In addition, organizations are motivated to adopt innovation that their competitors have adopted [16]. Hence, Competition Intensity has been considered as an adoption determinant for Blockchain including Permissioned Blockchain [35].

6. Insights on Permissioned Blockchain Adoption

6.1 Background of Executive Interviews

Intending to ascertain the determinants listed in Section 5, we interviewed 10 senior IT executives in Hong Kong with good knowledge of Permissioned Blockchain. They came from different industries including Finance & Banking, Transportation, Public Utilities, Information Technology, to Retail that can provide broad perspectives.

IT executives were interviewed because they are the earliest ones in their organizations to explore innovations and evaluate the viability of adoption. They are also technically savvy to understand Permissioned Blockchain and able to provide professional insights.

Hong Kong-based executives were selected because Blockchain technology has been growing rapidly in Hong Kong. According to a report from the Financial Services and Treasury Bureau of the Hong Kong Government, there were 57 Fintech firms established in Hong Kong in 2019. 45% of them are developing Blockchain applications for large businesses.[14]

[14] Source: https://www.crowdfundinsider.com/2020/06/162413-blockchain-was-ranked-largest-fintech-sector-in-hong-kong-last-year-report/.

Table 3. Profile of permissioned blockchain executives interviewed.

	Alias	**Industry**	**Level of Seniority in IT**	**Number of employees**	**Corporate Nature**	**Adopter/Non-Adopter**	**Individual/ Consortium**
1	AI-1	Travel	Top	< 500	Privately Held	Adopter	Individual
2	AI-2	Transport	Top	1,001–5,000	Public Company	Adopter	Individual
3	AI-3	Financial Services	Top	< 500	Privately Held	Adopter	Individual
4	AC-1	Banking	Top	> 10,000	Public Company	Adopter	Consortium
5	AC-2	Banking	Senior	> 10,000	Public Company	Adopter	Consortium
6	AC-3	IT & Services	Senior	> 10,000	Public Company	Adopter	Consortium
7	AC-4	Transport	Top	> 10,000	Privately Held	Adopter	Consortium
8	N-1	Utilities	Top	5,001–10,000	Public Company	Non-Adopter	
9	N-2	Transport	Top	> 10,000	Public Company	Non-Adopter	
10	N-3	Retail	Top	> 10,000	Privately Held	Non-Adopter	

One of the reasons for the fast growth of Blockchain companies in Hong Kong is its strategic position as a digital technology hub in Asia.

Both adopters and non-adopters were interviewed to secure a balanced view. For adopters, both individual adopters and consortium adopters were interviewed to understand the drivers behind the selection. Non-adopters interviewed all had good knowledge of Permissioned Blockchain and their decisions of not adopting at this stage were informed decisions.

These interviews generated valuable insights for the further understanding of the adoption of Permissioned Blockchain. The below table listed the profile of the interviewees.

6.2 Senior IT Executives Insights on Permission Blockchain Adoption Determinants

Perceived Benefits are important but Top Management Support sometimes is even more important as a determinant

Perceived Benefits are one of the most important determinants for any corporate purchase decision as mentioned in previous innovation adoption literature [30,43]. During the executive interviews, almost all executives agreed that Perceived Benefits should be a factor to examine as an adoption determinant but not necessarily the most important factor at all times. One individual adopter commented that Perceived Benefits are important when the cost and benefits are more quantifiable. That will be in at or beyond the growth stage in the typical product life cycle when there are references that can be drawn from other organizations' adoption. However, it is not in the development or introduction stage of Permissioned Blockchain. During the early stages of the lifecycle, adopters may just want to be an innovator to be the "first mover" before seeing concrete benefits [37].

Therefore, Top Management Support is a crucial adoption determinant of Permissioned Blockchain during the early stages when the benefits are uncertain, and the risk of failure is high. According to the individual adopter from the transportation industry interviewed, he would not proceed to adopt Permissioned Blockchain if he has not gotten the commitment from his top management in terms of endorsing its strategic importance to the whole organization, investing in the resources to adopt, and willing to take risks for failure. His viewpoint was echoed by other adopters and also non-adopters. One non-adopter, also from the transportation industry, commented that a key reason for his non-adoption is his top management placed higher priorities on an optimal service operation and do not want to take the risk of adopting Permissioned Blockchain in its supply chain processes. His top management's fear is if the suppliers cannot comply with the adoption and lead to an interruption of spare parts supply, which eventually may impact its service operation.

The criticality of trading partner readiness on adoption depends on the complication of business cases, the number of trading partners, and the changes in the trading partners required.

Implementation of Permissioned Blockchain as IOS is for external processes involving trading partners. For supply chain digital transformation, these trading partners are suppliers, distributors, or customers, as well as service providers such as banks or logistics companies. All the adopters interviewed were transacting with multiple external trading partners but had different views on the importance of trading partner readiness on adoption.

The CIO deployed Permissioned Blockchain on the marketing loyalty program felt that TP Readiness is not critical for adoption. It is because he believed that Permissioned Blockchain technology such as Decentralized Applications (DApp) makes trading partner onboarding not difficult even for non-technology savvy trading partners. They just needed to use the application to input or extract data. They did not even need to know Permissioned Blockchain technology for supporting the adoption. Therefore, TP Readiness is less critical for simple business cases which involve limited business change and simple technical enablement for the trading partners.

According to a CIO in the transportation industry who is an individual adopter, the criticality of trading partner readiness is even lower if the organization has negotiation power over its trading partners, i.e., if it is a strategic buyer or supplier. It is because the trading partner will have to get itself ready to retain the business relationship with the organization.

Another CIO from a public utility, who is a non-adopter felt that trading partner readiness should not only refer to the readiness on the technical side but the business side for orchestrating a new inter-organizational process as well. He stressed that the trading partners have to see benefits so that they will put the resources and effort to cope with the Permissioned Blockchain adoption. He used his multi-tier supply chain ecosystems that involve a large number of trading partners to elaborate on the importance of trading partner readiness. In his case, the readiness of some trading partners is not sufficient for adopting Permissioned Blockchain in a use case that involved all the trading partners. For instance, the track and chase of spare parts for maintenance. His company used different suppliers to provide the same spare parts for the maintenance of its equipment. One can see that if Permissioned Blockchain cannot be implemented for all suppliers, it will make his situation even more complicated. It is because if some suppliers are ready with one Permissioned Blockchain-enabled spare parts ordering process and the others are not, his organization still has to maintain dual processes in its supply chain for spare parts tracking which does not improve its efficiency and effectiveness. Therefore, the more complicated the business process, the more trading partners, and the more changes required on the trading partner side, the more critical a trading partner's readiness for Permissioned Blockchain adoption.

Individual-firm Adoption and Adoption via Blockchain Industry Consortium are not mutually exclusive and can co-exist.

As mentioned in Section 4, there are two adoption models of Permissioned Blockchain including individual-firm adoption and adoption via a Blockchain Industry Consortium model. The interviews with IT executives illustrated that organizations may adopt multiple Permissioned Blockchains for different business purposes. Hence, they may adopt both individually and via a Blockchain Industry Consortium moel.

A CIO of a global financial institution adopted both forms of Permissioned Blockchain. His individual-firm adoption aimed at improving inter-organizational business processes with trading partners in the mortgage loan application process. The focus is the immutability of the historical transaction data of the property as well as the corresponding documents of the transactions including the agreements for sales and purchase. Key data were kept in the Permissioned Blockchain for mortgage loan applications. By doing so, the financial institution did not require to conduct further validation on the documents manually for the application which saved its costs.

Organizational Readiness is an outcome of Top Management Support

Some executives interviewed highlighted the importance of an organizational readiness in adopting Permissioned Blockchain. As highlighted by the CIO from the transportation industry as an adopter, an organization needs to be ready when it decides to adopt Permissioned Blockchain. Such readiness includes the availability of financial resources as well as non-financial resources such as talents in Permissioned Blockchain technology and business transformation. These are crucial for the organization to adopt Permissioned Blockchain successfully. Therefore, Organizational Readiness was named as a possible adoption determinant.

From an innovation adoption perspective, Organizational Readiness can be assessed by the availability of financial resources and non-financial resources such as IT resources [44]. In our discussion with other interviewees, a few of them pointed out that resources are not a showstopper if the adoption has the support of the top management. Top Management Support, as mentioned in Section 5.2.2, includes the commitments of the organization's leaders in allocating resources and taking risks [30]. Therefore, these executives agreed that Organization Readiness is an outcome of Top Management Support instead of an individual-firm adoption determinant of Permissioned Blockchain. They concurred on the importance of top management support.

Non-adoption is due to the lack of adoption determinants

There were three non-adopters we interviewed. All of them agreed that their non-adoption was due to the lack of certain adoption determinants. The CIO of a retail group advised that he had not identified use cases that demonstrated benefits significant enough to justify the investment. There was also no relevant blockchain industry consortium in the industry his company operates. He has been explored with other industry players the idea of forming an industry consortium, but it was not successful. Hence, he was still in a status quo position.

The CIO of the public utility company cited similar reasons. He has been observing Permissioned Blockchain adoption in the energy industry but still not able to identify the one which he expected can bring sufficient benefits to justify. The public utility company offers services to a large community of customers. The most important consideration for them is to ensure the quality and stability of their services to meet the public's expectations. On the other hand, they have not seen an urgent need for adopting innovation aggressively for a competitive reason as there were few competitors in their sector. From these two perspectives, they were taking a conservative approach in their Permissioned Blockchain adoption strategy. The CIO also did not consider Permissioned Blockchain technology mature enough as there was a lack of proven success in adopting Permissioned Blockchain in his sector to serve a large user community. The CIO also did not see all the trading partners were ready as mentioned above in the point regarding trading partner readiness.

Another non-adopter, the CIO of the transportation company, agreed that the lack of readiness of all trading partners is a key inhibitor for them to adopt Permissioned Blockchain in their supply chain. Similar to his counterpart in the utility business, he saw the top management's focus as the safety and stability of services. Adopting Permissioned Blockchain is not a key priority of the top management. Rather their top management felt that it was more practical to adopt technologies for improving cost efficiency directly such as Robotic Process Automation (RPA).

7. Conclusion

Permissioned Blockchains are an important enabler for supply chain digital transformation given its technical characteristics and business models supported. While Permissioned Blockchains supports the closed user group concept as well as the business models of IOS, it enables a new, distributed model for mission-critical business transactions between trading partners with a low level of trust. This enablement opens more possibilities for IOS in supply chain digitization such as the bundling with IoT for the autonomous supply chain. In addition, the Blockchain Industry Consortium model is an effective adoption model for minimizing adoption risks and maximizing benefits in addressing industry-wide business pain points that adopters or potential adopters should consider for expediting adoption. Unlike Permissionless Blockchains, there are technological, organizational, inter-organizational, and environmental determinants for Permissioned Blockchains as adopting it comes down to a corporate decision making process. It is crucial to understand these determinants for any organization in making an informed adoption decisions. Senior IT executives are the earliest ones in an organization to identify new technologies and making adoption decisions. Their opinions are therefore valuable. While some of those are shared in this chapter, readers are encouraged continuously to seek senior IT executives' advice on the further development of Permissioned Blockchains and related technologies, on the viability of adoption.

Acknowledgements

The writer would like to acknowledge Prof. Yulin Fang, University of Hong Kong, and Dr. Ting Xu, Xi'an Jiaotong University, for their advice in reviewing this chapter. The 10 senior IT executives are also acknowledged for their valuable input on the Permissioned Blockchain adoption determinants. This chapter would not be a completed one without their insights.

Glossary

Permissionless Blockchain: Also named as public blockchain. It is the type of blockchain open for anyone with anonymity of each's identify to participate.

Permissioned Blockchain: It is the type of blockchain only be accessed by pre-approved participants. It includes both federated/consortium blockchain and private blockchain.

Inter-organizational Systems: Information systems used by organizations and their business partners for exchanging business transaction data and orchestrating business processes.

Blockchain Industry Consortium: A business model for players from the same industry ecosystem to adopt permissioned blockchain collectively to operate the blockchain governance. They also enjoy the benefits of sharing the cost and risk and reaching to the critical mass quicker.

TOE Framework: A theory studying determinants of innovation adoption by enterprise. Determinants are categorized into technological, organizational, and environmental.

References

[1] Nakamoto, S. 2008. Bitcoin: A Peer-to-Peer Electronic Cash System. *Www.Bitcoin.Org*, pp. 1–29.

[2] Zheng, Z., Xie, S., Dai, H., Chen, X. and Wang, H. 2017. An overview of blockchain technology: architecture, consensus, and future trends. *Proc. - 2017 IEEE 6th Int. Congr. Big Data, BigData Congr. 2017*, no. June, pp. 557–564.

[3] Helliar, C. V., Crawford, L., Rocca, L., Teodori, C. and Veneziani, M. 2019. Permissionless and permissioned blockchain diffusion. *Int. J. Inf. Manage.*, vol. 54, no. October 2019, p. 102136, 2020.

[4] Rangaswami, J., Warren, S., Mulligan, C. and Zhu Scott, J. 2018. Blockchain beyond the hype a practical framework for business leaders. *White Pap. World Econ. Forum 2018*, no. April, 2018.

[5] Singh, S. K., Jenamani, M., Dasgupta, D. and Das, S. 2021. A conceptual model for Indian public distribution system using consortium blockchain with on-chain and off-chain trusted data. *Inf. Technol. Dev.*, 27(3) 499–523.

[6] Li, K., Lee, J. Y. and Gharehgozli, A. 2021. Blockchain in food supply chains: a literature review and synthesis analysis of platforms, benefits and challenges. *Int. J. Prod. Res.*.

[7] Angelis, J. and Ribeiro da Silva, E. 2019. Blockchain adoption: A value driver perspective. *Bus. Horiz.*, 62(3): 307–314.

[8] Tapscott, D. and Taoscott, A. 2016. *Blockchain Revolution: How the Technology Behind Bitcoin Is Changing Money, Business, and the World*. New York: Portfolio.

[9] Viriyasitavat, W. and Hoonsopon, D. 2019. Blockchain characteristics and consensus in modern business processes. *J. Ind. Inf. Integr.*, 13, no. June 2018, pp. 32–39.

[10] Helliar, C. V., Crawford, L., Rocca, L., Teodori, C. and Veneziani, M. 2020. Permissionless and permissioned blockchain diffusion. *Int. J. Inf. Manage.*, vol. 54, no. April, p. 102136, 2020.

[11] Kuan, K. K. Y. and Chau, P. Y. K. 2001. A perception-based model for EDI adoption in small businesses using a technology-organization-environment framework. *Inf. Manag.*, 38(8): 507–521.

[12] Cash, J. I. J. and Konsynski, B. R. 1985. IS redraws Competitive Boundaries. *Harvard Bus. Rev. HBR*, no. March-April, pp. 134–142.

[13] Johnston, H. R. and Vitale, M. R. 1988. Creating competitive advantage with interorganizational information systems. *MIS Q.*, 12(2): 153.

[14] Lyytinen, K. and Damsgaard, J. 2011. Inter-organizational information systems adoption-a configuration analysis approach. *Eur. J. Inf. Syst.*, 20(5): 496–509.

[15] Iacovou, C. L., Benbasat, I. and Dexter, A. S. 1995. Electronic data interchange and small businesses: adoption and impact of technology. *MIS Q.*, 19(4): 465–485.

[16] Teo, H. H., Wei, K. K. and Benbasat, I. 2003. Predicting intention to adopt interorganizational linkages: An institutional perspective. *MIS Q. Manag. Inf. Syst.*, 27(1): 19–49.

[17] Toppen, R., Smits, M. and Ribbers, P. 1998. Financial securities transactions: A study of logistic process performance improvements. *J. Strateg. Inf. Syst.*, 7(3): 199–216.

[18] Sawik, T. 2017. A portfolio approach to supply chain disruption management. *Int. J. Prod. Res.*, 55(7): 1970–1991.

[19] Mendling, J. et al. 2018. Blockchains for business process management - challenges and opportunities. *ACM Trans. Manag. Inf. Syst.*, 9(1): 31–35.

[20] Werner, F., Basalla, M., Schneider, J., Hayes, D. and Vom Brocke, J. 2020. Blockchain adoption from an interorganizational systems perspective–a mixed-methods approach. *Inf. Syst. Manag.*, 38(2): 135–150.

[21] Pun, H., Swaminathan, J. M. and Hou, P. 2021. Blockchain adoption for combating deceptive counterfeits. *Prod. Oper. Manag.*, 30(4): 864–882.

[22] Wang, Y., Singgih, M., Wang, J. and Rit, M. 2019. Making sense of blockchain technology: How will it transform supply chains? *Int. J. Prod. Econ.*, 211: 221–236.

[23] Mitra, S. and Singhal, V. 2008. Supply chain integration and shareholder value: Evidence from consortium based industry exchanges. *J. Oper. Manag.*, 26(1): 96–114.

[24] Zhao, K., Xia, M. and Shaw, M. J. 2007. An integrated model of consortium-based e-business standardization: Collaborative development and adoption with network externalities. *J. Manag. Inf. Syst.*, 23(4): 247–271.

[25] Zavolokina, L., Ziolkowski, R., Bauer, I. and Schwabe, G. 2020. Management, governance, and value creation in a blockchain consortium. *MIS Q. Exec.*, 19(1): 1–17.

[26] Furlonger, D. and Kandaswamy, R. 2019. Hype cycle for blockchain business. *Gart. Inc.*, no. July, p. 77.

[27] Bengtsson, M. 2000. 'Coopetition' in Business Networks - to Cooperate and Compete Simultaneously. *Ind. Mark. Manag.*, 426: 411–426.

[28] Tornatzky, L. G., Fleischer, M. and Chakrabarti, A. 1990. *The Process of Technological Innovation*. Lexington, Mass.: Lexington Books.

[29] Pan, M.-J. and Jang, W.-Y. 2008. Determinants of the adoption of enterprise resource planning within the technology-organization-environment framework: Taiwan'S Communications Industry. *J. Comput. Inf. Syst.*, 48(3): 94–102.

[30] Lin, H. F. 2014. Understanding the determinants of electronic supply chain management system adoption: Using the technology-organization-environment framework. *Technol. Forecast. Soc. Change*, 86: 80–92.

[31] Chwelos, P., Benbasat, I. and Dexter, A. S. 2001. Research report: empirical test of an EDI adoption model. *Inf. Syst. Res.*, 12(3): 304–321.

[32] Teo, T. S. H., Lin, S. and hung Lai, K. 2009. Adopters and non-adopters of e-procurement in Singapore: An empirical study. *Omega*, 37(5):972–987.

[33] Tsou, H. T. and Hsu, S. H. Y. 2015. Performance effects of technology-organization-environment openness, service co-production, and digital-resource readiness: The case of the IT industry. *Int. J. Inf. Manage.*, 35(1): 1–14.

[34] Clohessy, T. and Godfrey, R. 2018. Organizational factors that influence blockchain adoption in Ireland. *Natl. Univ. Irel. Galw.*

[35] Wong, L. W., Leong, L. Y., Hew, J. J., Tan, G. W. H. and Ooi, K. B. 2019. Time to seize the digital evolution: Adoption of blockchain in operations and supply chain management among Malaysian SMEs. *Int. J. Inf. Manage.*, vol. 52, no. September 2019, p. 101997.

[36] Lee, J. Y., Swink, M. and Pandejpong, T. 2017. Team diversity and manufacturing process innovation performance: the moderating role of technology maturity. *Int. J. Prod. Res.*, 55(17): 4912–4930.

[37] Rogers, E. M. 1962. *Diffusion of innovations*, 3rd ed., vol. 17, no. 1. The Free Press.

[38] Lee, S. and jae Kim, K. 2007. Factors affecting the implementation success of Internet-based information systems. *Comput. Human Behav.*, 23(4): 1853–1880.

[39] Behnke, K. and Janssen, M. F. W. H. A. 2020. Boundary conditions for traceability in food supply chains using blockchain technology. *Int. J. Inf. Manage.*, vol. 52, no. March 2019, p. 101969, 2020.
[40] Lin, H. F. 2006. Interorganizational and organizational determinants of planning effectiveness for Internet-based interorganizational systems. *Inf. Manag.*, 43(4): 423–433.
[41] Zhu, K., Kraemer, K. and Xu, S. 2003. Electronic business adoption by European firms: A cross-country assessment of the facilitators and inhibitors. *Eur. J. Inf. Syst.*, 12(4): 251–268.
[42] Porter, M. E. and Millar, V. E. 1985. How information gives you competitive advantage. *Harvard Bus. Rev. HBR*, no. July, 1985.
[43] Wang, Y. S., Li, H. T., Li, C. R. and Zhang, D. Z. 2016. Factors affecting hotels' adoption of mobile reservation systems: A technology-organization-environment framework. *Tour. Manag.*, 53: 163–172.
[44] Iacovou, C. L., I. Benbasat and A. S. Dexter. 1995. Electronic data interchange and small organizations: Adoption and impact of technology. *MIS Q. Manag. Inf. Syst.*, 19(4): 465–485.

6

Blockchain, Supply Chain and Adoption

A Bibliometric Analysis

Colin Callinan, Dr. Amaya Vega, Dr. Trevor Clohessy and *Prof. Graham Heaslip*

1. Introduction

Blockchain can be defined as a 'peer-to-peer distributed data infrastructure' [1, p.1] that is able to 'store/record data and transactions backed by a cryptographic value' [2, p.1]. In laymen's terms, Blockchain enables a 'community of users to record transactions in a shared ledger within that community, such that under normal operation of the blockchain network no transaction can be changed once published' [3, p.iv]. The creation of Blockchain is credited to Satoshi Nakamoto, who in 2008 published an article describing a peer-to-peer electronic cash system that was to become Bitcoin [4]. This work built on previous concepts developed in the 1980s and 1990s by Leslie Lamport describing a consensus model built on a network of computers where the computers or network may be unreliable [3]. Nakamoto's article described, what was at that time, the current system of electronic currency transfer. One of the key inefficiencies identified was the mediation feature, exposed by the lack of trust inherent in the system. This system inefficiency adds additional transaction costs as it requires either a dispute resolution mechanism or the subsuming of costs associated with fraudulent behaviour thus rendering small/micro

Galway Mayo Institute of Technology.

transfers underutilised [4]. The publication of the Nakamoto article was serendipitous as it appeared at a time when trust in financial institutions had hit rock bottom caused by the advent of the 2008 financial crisis [5, p.11]. What followed was the operationalisation of the concept under the auspices of Bitcoin. A number of other networks have been created since, termed cryptocurrencies, describing 'mediums of exchange that uses cryptography to secure transactions-as against those systems where the transactions are channeled through a centralized trusted entity' [6, p.9]. There are now thousands of cryptocurrencies in existence with a market capitalisation of over 500 billion US dollars [7]. This new medium of exchange has implications for the traditional roles of intermediaries and institutions, possibly rendering them obsolete or, at the very least, reengineered. Purported changes include the restructuring of economic, financial, legal and governance systems [8].

Blockchain is a tamper proof distributed ledger technology (DLT) which is not susceptible to easy change. Its operation occurs in a shared environment where all the transactions are validated by users and are traceable [2]. This is done without a centralised intermediary. Users in the environment modify the ledger by executing pieces of code known as smart contracts. The smart contracts are then approved and ordered into a linked chain of blocks, (hence blockchain), distributed across multiple peers (i.e., users), ensuring the immutability of the data [9]. Blockchain technology appears deeply complex but can be understood by breaking it down to its' constituent parts. From a high level perspective, distributed ledger technology makes use of existing computer science mechanisms (e.g., private key storage), cryptographic algorithms (e.g., *Hashing* and asymmetric key cryptography), and record keeping concepts [3]. Private key storage refers to a form of cryptography that allows access to a digital asset, e.g., a cryptocurrency. Private keys are usually stored using a piece of software known as a wallet. This is a crucial aspect of Blockchain technology as the digital asset is associated with the key. If the key is lost or stolen then the digital asset is also lost as it is computationally impractical to regenerate the same key [3]. Hashing is a function that meets the encryption mandate required to solve a blockchain computation. Users can take input data, of almost any size, hash the data and obtain the same result thereby proving there was no change in the data. The security advantage of this method is that even the smallest change in the data will generate a different output [3]. Asymmetric key cryptography refers to public key cryptography. Both private keys (discussed above) and public keys are utilised and are mathematically related to each other. This function is in essence the trust feature in-built into blockchain. The user's wallet public key acts as the address which is visible to everyone while private keys are secret and used to digitally sign transactions. The asymmetric key cryptography facilitates the trust relationship between actors who do not know or trust one another by allowing the a means to verify the authenticity of transactions which remain public [3,10]. The aspect of record keeping concepts that we are concerned with here is the ledger aspect of blockchain technology. Traditionally, a ledger is a collection of transactions that usually took the form of pen and paper. More recently, these ledgers have been stored digitally primarily in databases operated by the ledger owner on behalf of clients/customers, etc. The key difference with distributed ledger technology is that control does not lie with one actor but with several or all network participants. This differentiates it from cloud computing and data replication as these technologies employ existing shared ledgers. This affords distributed ledger technology its immutable characteristic—no single actor can amend or approve new additions to the ledger [3,11].

A supply chain can be defined as a complex adaptive system network that traverses multiple stages, relationships, geographical locations, various financial systems, and multiple entities encapsulated by differing time-based pressures depending on the type of product and market [12,13]. Supply chains are typified by the inclusion of multiple partners. These may include but are not limited to manufacturing factories, distribution centres, suppliers, couriers, and ancillary logistic services [14]. As such, supply chains are increasingly becoming more complex. A number of factors have contributed to this—the search for sustainability, increasing globalisation, trade liberalisation, reduction in trade costs and the application of new technologies. In addition, the management of supply chain networks is a crucial factor in preserving organisational competitiveness [12,15]. Another factor in the monitoring and transportation of goods is that of a supply chain management system. This can be defined as 'an enterprise system that supports or manage supply chain activities and integrated management process. SCMS manages several business entities such as suppliers, retailers, manufacturers, and distributors who process raw materials processed into a product and then distributed to retailers' [16, p.1]. Blockchains role within supply chain has gained increasing attention from researchers and practitioners in recent times. A number of purported benefits have been listed including smart contracts, product traceability, enforcement tracking, stock control, transaction and settlement, and information immutability [17].

The pace of technological innovation and change has necessitated the speedy adoption of information communication technologies as a crucial goal for firms. This adoption/non adoption affects many facets of an organisation including attaining competitive advantage, increased revenue, reduction of operational costs, and improving operational efficiency [18]. It is therefore crucial that the factors that determine the adoption of information technology at the firm level are clearly understood. Many theories have been posited in the past, however, the majority view the issue of technology adoption occurs at the individual level of analysis. The two dominant theories of information technology adoption emanating from information systems discipline that employ a firm level of analysis, are the diffusion of innovation theory and the technology, organisation, and environment framework. The diffusion of innovation (DOI) theory, posited by Rogers, describes how, and why the rate at which new ideas and technology spread through culture at both individual and firm level. Organisational innovativeness is determined by individual characteristics, such as attitude towards change, internal characteristics of the organisational structure, such as complexity and interconnectedness, and finally the external characteristics of the organisation such as system openness [19]. The technology, organisation and environment (TOE) model was first developed in 1990 by [20]. It pinpoints three distinct areas of an organisations context that affects the process of adoption and implementation of a technological innovation: the technological context, organisational context and environmental context [19]. The technological context refers to both the internal and external technologies pertinent to the firm. From a more specific perspective, the technological context incorporates factors such as complexity, relative advantage, privacy, security, and compatibility. These factors have been shown previously to affect existing or potential information technology adoptions [19,21,22]. An abstract view of the organisational context describes the firm by examining its scope, size, and managerial structure. More specific considerations include; top management support, prior IT experience, innovativeness, information intensity, and organisational readiness [19,21,23]. The environmental context refers to the wider area in which a company operates its business, i.e., the industry, its competitors and relationships with government including regulations [19,21].

The remainder of this chapter is organised as follows; an overview of related work and previous publications that have examined both bibliometric analyses of blockchain in a number of different domains and bibliometric analyses of blockchain in supply chain. Following this, the methodology is described. A descriptive analysis and network analysis of the data is then presented with accompanying visualisations. The chapter ends with a discussion and conclusion.

2. Related Work

In recent years, various approaches for evaluating publications have been adopted. Bibliometric and citation analysis is well-established as a study assessment tool, especially when it comes to analysing the content and network of publications. Bibliometric analysis is a form of quantitative analysis used to evaluate scientific publications. This approach simplifies authorship and keyword analysis by focusing on many factors such as authorship and keywords and terms [24]. Bibliometric analysis aids in the holistic comprehension of a study topic, as well as the mapping of its borders, in addition to the discovery of significant authors and new research avenues [25]. It further facilitates the comprehension of research activities' qualities, organization, and patterns [26].

This approach has previously been used frequently by many authors in a number of disciplines. From the accounting domain [27], mapped the present level of knowledge regarding blockchain technology in the accounting area, as well as identifying issues and problems for future research. Their focus was primarily identifying the important research destinations at the interface of blockchain and accounting. In higher education [28], examined the most common applications of blockchain technology in higher education in order to add to the body of knowledge in this sector. As a result, they were able to identify three strategies for advancing this field of study, engineering education, blockchain, and artificial intelligence. The overlap of blockchain and the internet of things was surveyed by [29], where they identified rapidly growing interest in this research area since 2017, further identifying computer science, telecommunications, engineering and communications as the most examined research areas. The application of blockchain technology in the food and agriculture sector was investigated by [30]. They found increasing publications from 2019 showing that traceability, supply chain, IoT, smart contract, and food safety are the top keywords that have occurred most frequently in this domain. From these keywords, three key research clusters emerged; the traceability system, the technological issues of blockchain, and the research development and benefits of the blockchain. Bibliometric analysis of blockchain and smart contracts was conducted by [31]. The research found that there were six leading streams that are currently appearing in literature; basic research on blockchain and smart contracts, applications of smart contracts to the internet of things, how public blockchain infrastructure Ethereum smart contracts can be standardized, secured, and verified, social and economic aspects and implications of smart contracts, challenges and potentials of smart contracts and focuses on use cases, and legal issues. A bibliometric analysis of the research focused on applying blockchain in management and its related areas performed by [25] led to the identification of a number of research themes in this area; strategic and regulatory issues that affect bitcoin and blockchain implementation, the unique characteristics of blockchain, their applicability for managerial processes, and the advantages and challenges of blockchain's integration within a business's framework, multi-domain deployment and understanding the application of blockchain in terms of financial markets, specifically, its original inception as the foundational technology for bitcoin. Bibliometric analysis of blockchain applications in power systems completed by [26] examined the research activity by country, the key research areas, the leading institutions, authors, publishers, most cited articles and keyword frequency. They found that authors from the USA and China were leading research in this area and that the preponderance of the publications come from the computer science and engineering domains.

Bibliometric analyses of blockchain and supply chain are negligible but are gaining increasing attention from researchers. An article from [32] explored blockchain technology in logistics and supply chain management finding five different research clusters in this area; theoretical sensemaking, conceptualizing and testing blockchain applications, framing blockchain technology into supply chains, the technical design of blockchain technology applications for real-world logistics/supply chain management applications, and the role of blockchain technology within digital supply chains. The role of blockchain in supply chain management was examined by [24] enabling them to detect the key supply chain domains and other integrated technologies. They identified a number of key areas in supply chain where blockchain could be considered a value add; management of supply chain, finance, logistic, risk, information management, and manufacturing, smart contracts, and the internet of things. Blockchain technology in food supply chains was examined by [33]. They found researchers have primarily focused on the applicability of blockchain for product authentication, finance, and logistics. In addition, 'blockchain is likely to become a dominant technology for enhancing transparency and traceability, reducing risk and, most importantly, enhancing trust among different stakeholders in the area of food supply chains' [33, p.1]. A further study of blockchain technologies in logistics and supply chain management was conducted by [34]. They found that 'the literature concentrates mainly on the conceptualization of blockchain; its potentials for supply chain sustainability; its adoption triggers and barriers; and its role in supporting supply chain agility, trust, protection of intellectual property, and food/perishable supply chains' [34, p.1].

3. Methodology

VOSviewer (https://www.vosviewer.com/) is a computer program developed by [35], and used primarily for building and viewing bibliographic maps. It has two key features that are useful to academics seeking to visualise relationships in bibliographic data—maps of authors or journals can be created based on co-citation data and/or maps of keywords based on co-occurrence data [35]. Bibliometric networks have been extensively studied for decades within the field of bibliometrics. However, one area that has been neglected somewhat is that of methodology, i.e., the actual construction of bibliographic networks. The central problem with bibliographic construction of maps is the assignment of co-authored publications to individual authors [36]. The common approach is what is known as the full counting method where a publication has been co-authored by three researchers, for example, and each author has been assigned a weight of one. The key advantage of VOSviewer is the employment of what is known as fractional counting. This method applied to the preceding example would assign each author a fractional weight of 1/3. Thus, the total weighting of one is distributed across the individual co-authorship links equally. VOSviewer software works best when used to build bibliographic networks based on data obtained from Web of Science and Scopus [36]. This is the approach taken for this research as it is deemed a more accurate representation of the bibliographic networks. Web of Science is the oldest citation database in existence covering citation and bibliographic data beginning in 1900. Scopus covers a wider range of journals but is limited to articles published after 1995 [37]. This does not pose an issue for this research as the oldest article emanating from the literature review is Satoshi Nakamotos' article describing Bitcoin from 2008. Furthermore, the overwhelming majority of articles found in this review are post 2018. VOSviewer offers three different types of visualisations; Network, Overlay and Density visualisations. A sample of each will now be presented based on the search string—"blockchain" AND "supply chain" AND "adoption". This search string was chosen as it produced the largest result, thus making it the most appropriate example to visualise. This resulted in 307 records from Web of Science by searching all fields and 363 from Scopus by searching title, abstract and keywords. The different approaches were chosen to expand the Web of Science results and narrow the Scopus results. This was justified as the Scopus results for the all fields search resulted in a large quantity of irrelevant papers that did not address the chosen topics directly. The outcome of the descriptive and bibliometric analysis is presented below.

4. Analysis and Results

4.1 Descriptive Analysis

Data extracted from the Scopus and Web of Science databases was analysed in March 2022. For the purposes of analysis, we utilised both the in built visualisation tools provided by Scopus and Web of Science and Microsoft Power BI. Power BI is an interactive data visualisation software package developed by Microsoft with a principal focus on business intelligence. Figure 1 graphically represents the publication types emanating from the search string used on the Scopus database. The overwhelming majority of publications are journal articles with conference papers representing a significant minority.

Figure 2 graphically denotes the publications per year by the top five journals hosted on the Scopus database. It shows that blockchain adoption research along supply chain only began to gain prominence in 2018. The journals that have seen the most publications can be broadly grouped around production, computer science/engineering and sustainability pointing towards practical, operational, and sustainability concerns.

Figure 3 shows the results of the search string weighted by the subject area appearance. The most common subject area topics emanating from the search were business/management, computer science, engineering, and decision science.

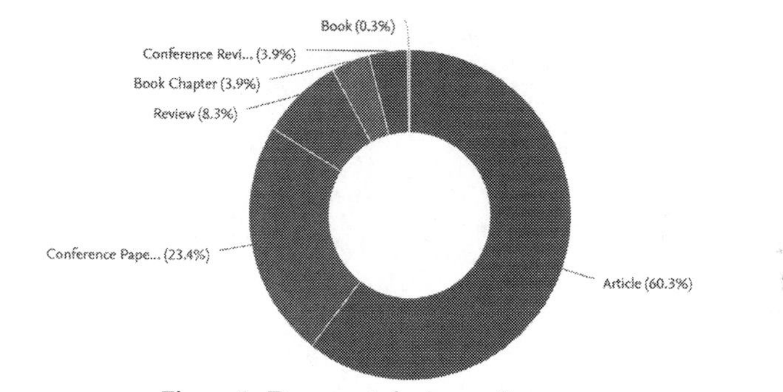

Figure 1. Documents by type—Scopus.

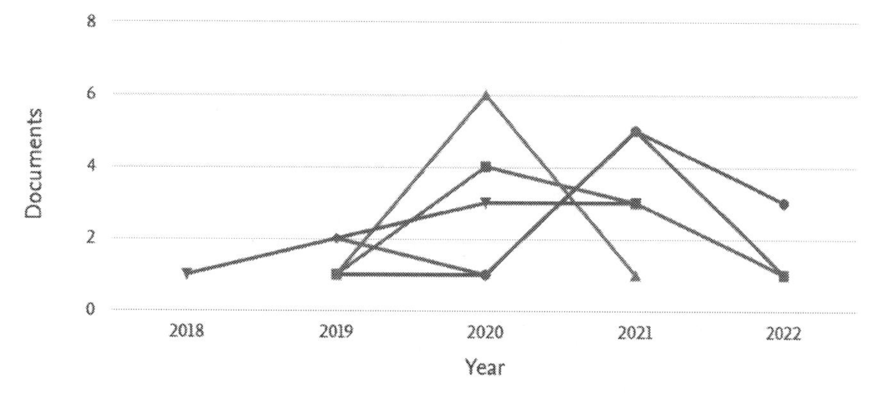

Figure 2. Documents per year by source—Scopus.

Figure 4 represents the top 15 journals hosted on the Scopus database by occurrence of publications based on the search string. The top 15 were selected as the remaining results give rise to negligible results. The majority of publications are concerned with production, operations, computer science, engineering, and logistics.

Table 1 and Figure 5 show the results of the search string weighted by the subject area appearance including individual count and percentage of total. The top 15 were chosen as the remainder of the categories gave rise to negligible results. The most common subject area topics emanating from the search were management, computer science and engineering.

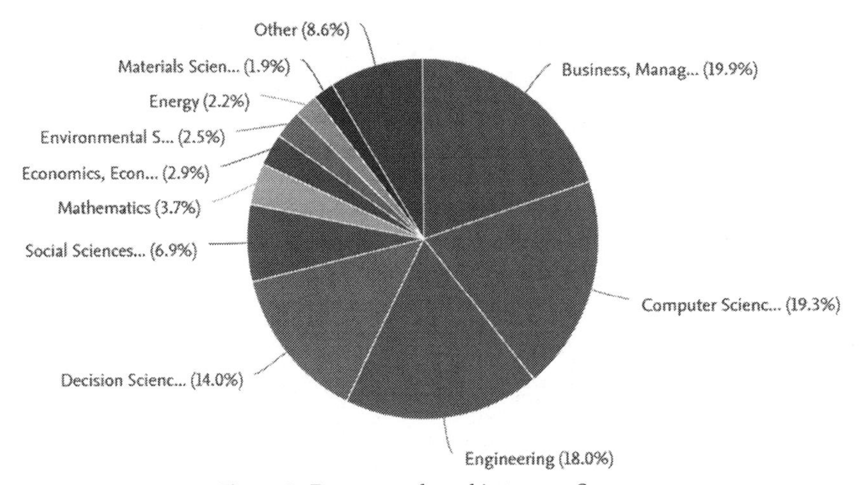

Figure 3. Documents by subject area—Scopus.

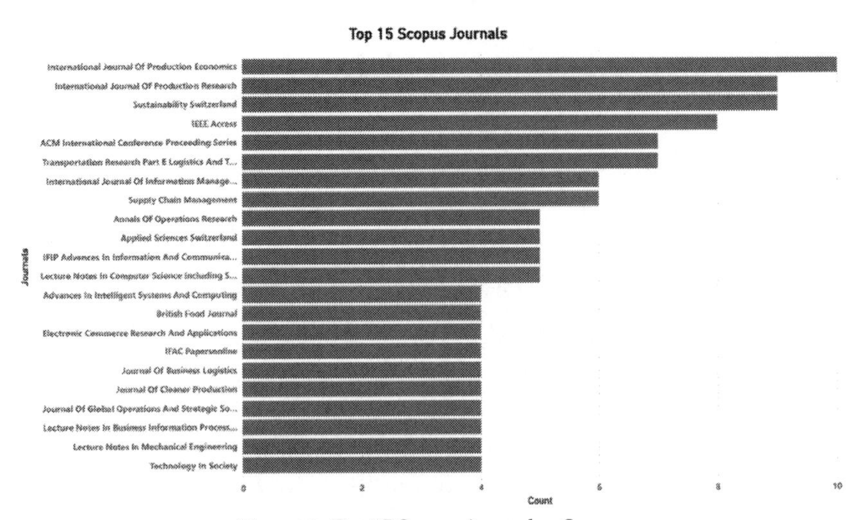

Figure 4. Top 15 Scopus journals—Scopus.

Table 1. Top 15 Web of Science categories.

Web of Science Categories	Record Count	% of Total
Management	90	29.316%
Operations Research Management Science	48	15.635%
Business	43	14.007%
Computer Science/Information Systems	42	13.681%
Engineering Industrial	37	12.052%
Computer Science Interdisciplinary Applications	30	9.7725
Engineering Electrical/Electronic	24	7.818%
Engineering Manufacturing	23	7.492%
Information Science	23	7.492%
Environmental Sciences	22	7.166%
Telecommunications	21	6.840%
Green Sustainable Science/Technology	20	6.515%
Environmental Studies	15	4.886%
Economics	13	4.235%
Engineering Civil	11	3.583%

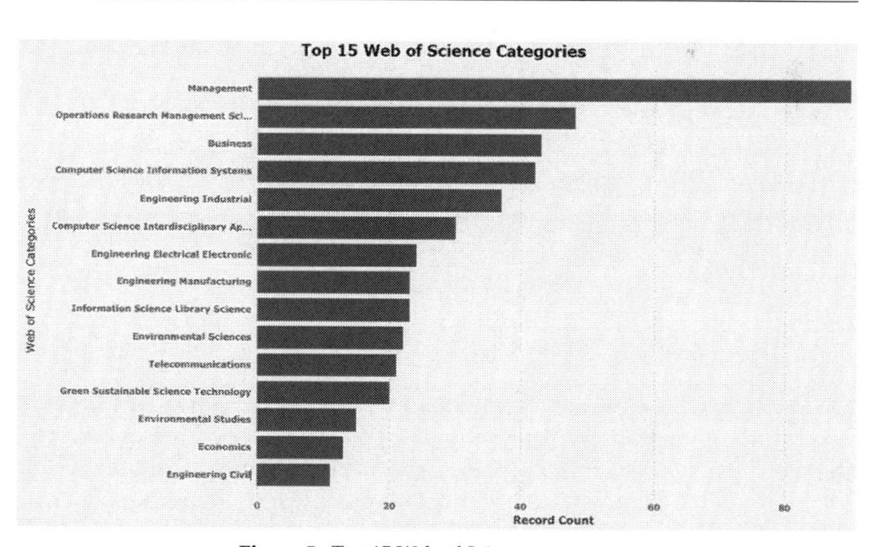

Figure 5. Top 15 Web of Science categories.

4.2 Network Analysis

Network Visualisations—Bibliographic & Textual Data

Items are represented by a label. In the following examples, the label is a term emanating from the titles and abstracts (keywords) or the text of the publications and a circle. The size of the label/term and circle is determined by the weighting of that item. The higher the weighting, the larger the label and circle. The terms are located based on the co-occurrence in the titles and abstracts, the higher the number of co-occurrences, the closer they will be to each other [38]. Figure 6 represents the co-occurrence of keywords from the Scopus database. Figure 7 represents the co-occurrence of keywords from the Web of Science database. Figure 8 represents the co-occurrence of terms from the Scopus database. Figure 9 represents the co-occurrence of terms from the Web of Science database. Colours indicate clusters of related terms.

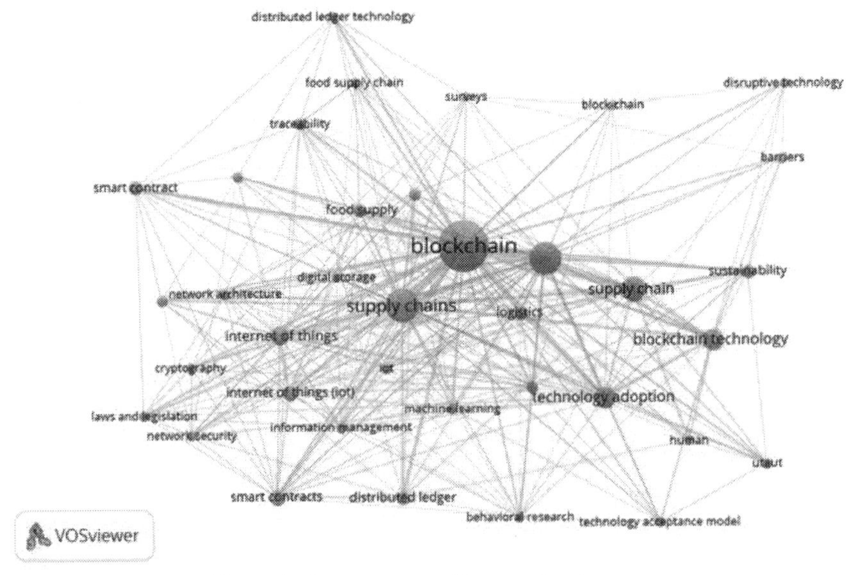

Figure 6. Co-occurrence of keywords—Scopus.

Overlay Visualisations – Bibliographic & Textual Data

Overlay visualisations represent the same data as the network visualisations except they are coloured differently. The colours assigned are denoted by the score given, with blue representing the lowest score and green/yellow the highest [38]. Figure 10 represents the co-occurrence of keywords from the Scopus database. Figure 11 represents the co-occurrence of keywords from the Web of Science database. Figure 12 represents the co-occurrence of terms from the Scopus database. Figure 13 represents the co-occurrence of terms from the Web of Science database.

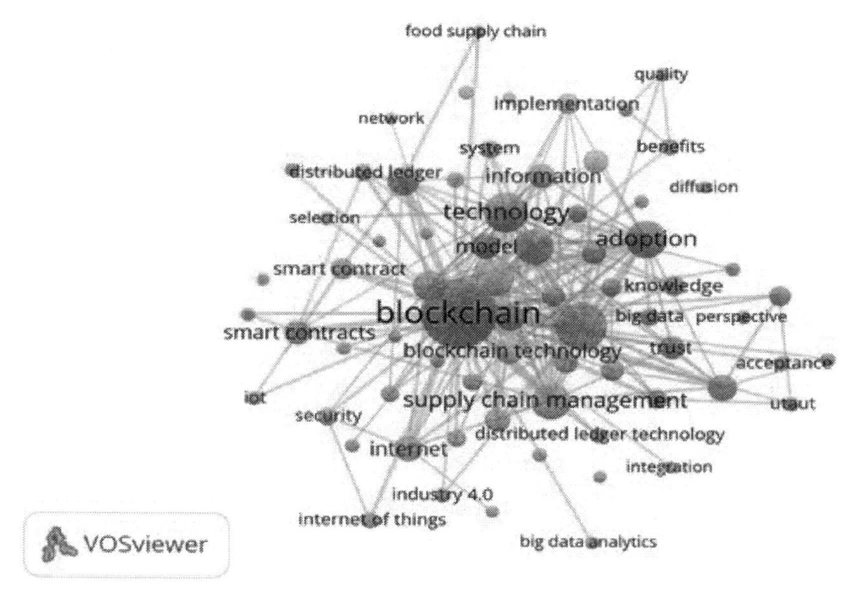

Figure 7. Co-occurrence of keywords—Web of Science.

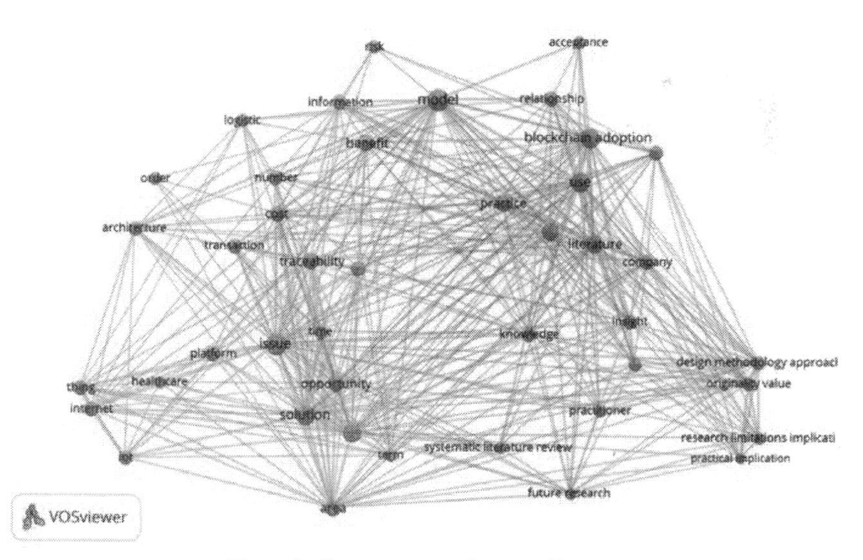

Figure 8. Co-occurrence of terms—Scopus.

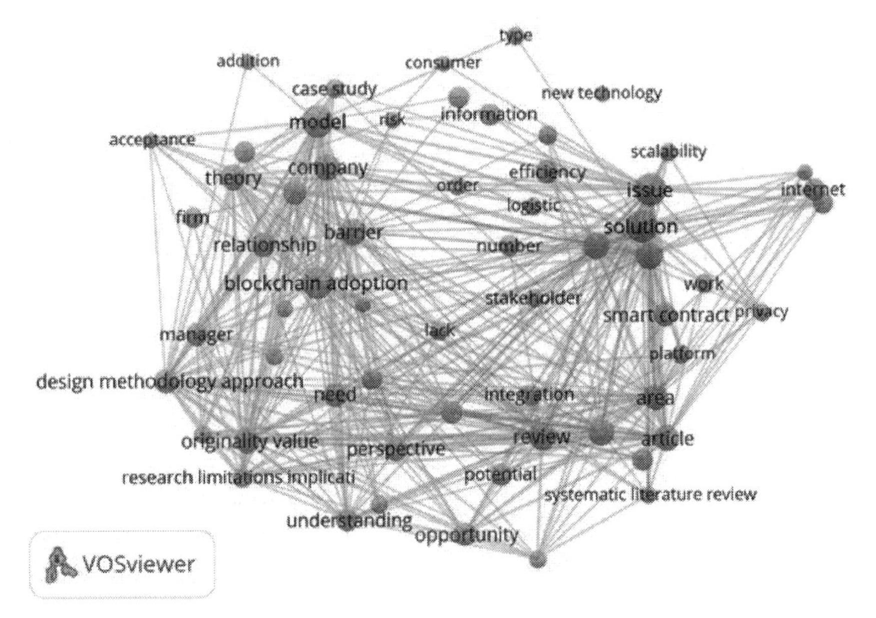

Figure 9. Co-occurrence of terms—Web of Science.

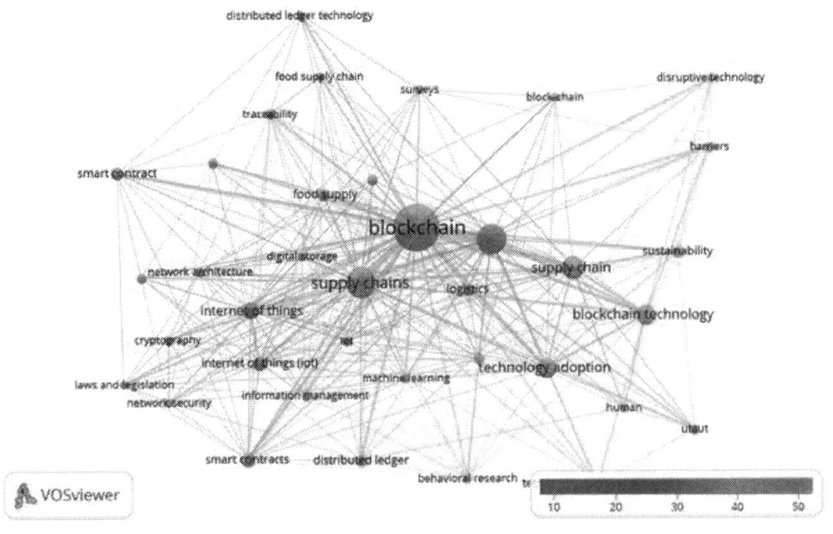

Figure 10. Co-occurrence of keywords—Scopus.

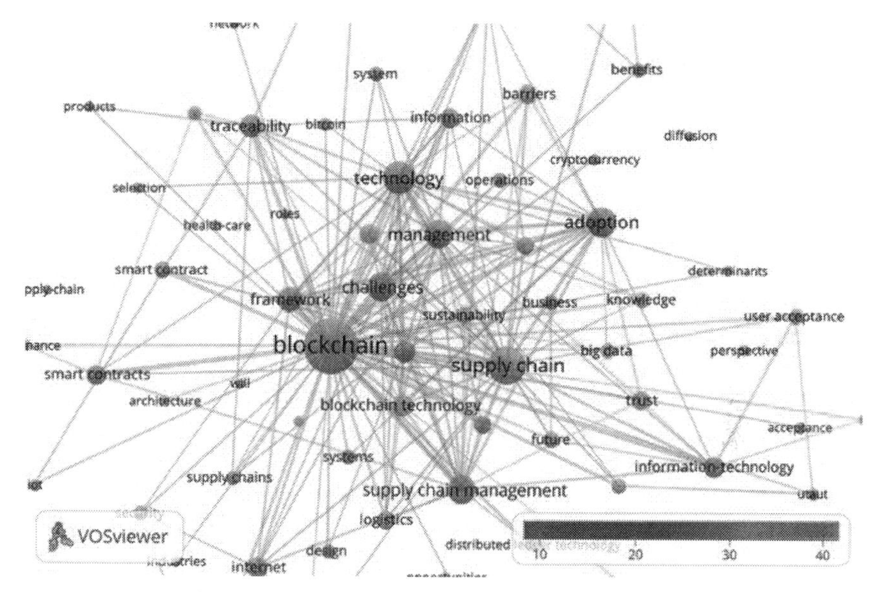

Figure 11. Co-occurrence of keywords—Web of Science.

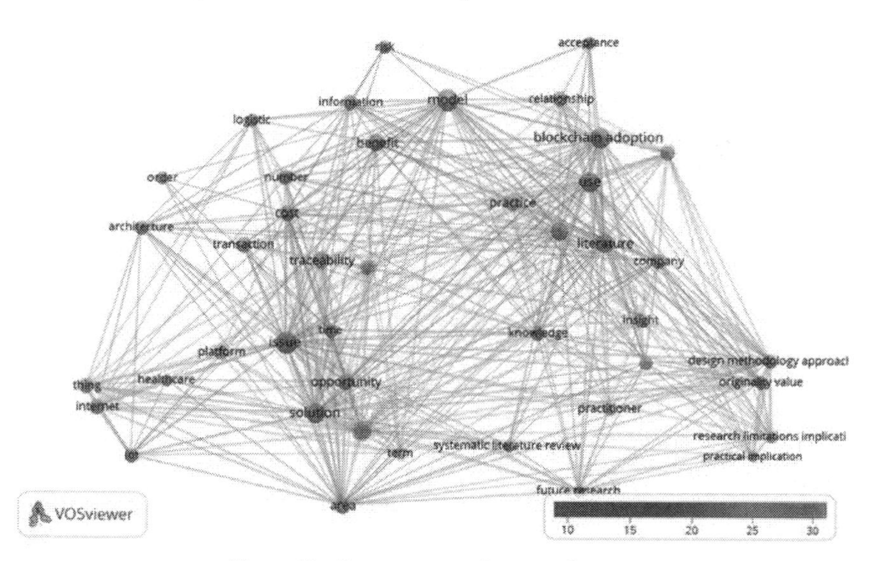

Figure 12. Co-occurrence of terms—Scopus.

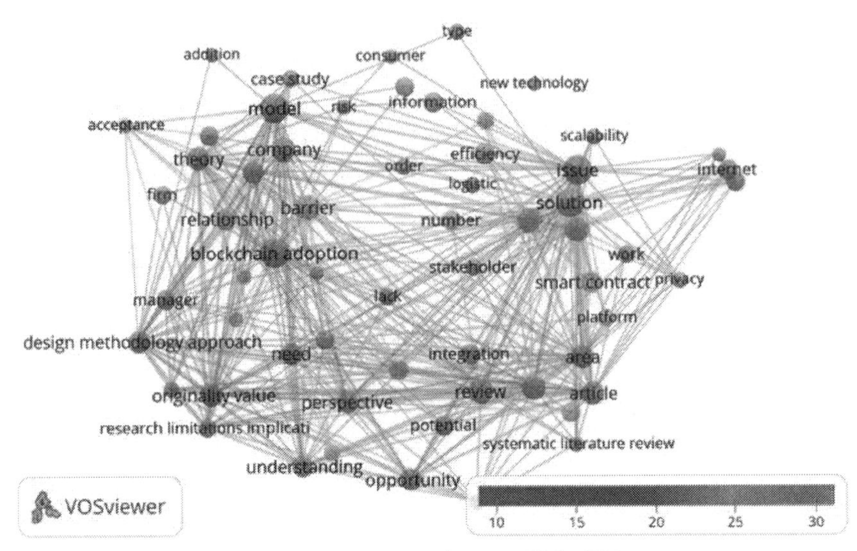

Figure 13. Co-occurrence of terms—Web of Science.

Density Visualisations—Bibliographic & Textual Data

The density visualisation represents items as labels in the same way as the network and overlay visualisations. The colour in the visualisations denotes the density of the items at that point. Similar to the overlay visualisations, colours range from blue to green to yellow. The larger the number of items in the area of a point and the higher the weights of the neighbouring items, the closer the colour of the point is to yellow. In contrast, the smaller the number of items in the area of a point and the lower the weights of the neighbouring items, the closer the colour of the point is to blue [38]. Figure 14 represents the co-occurrence of keywords from the Scopus database. Figure 15 represents the co-occurrence of keywords from the Web of Science database. Figure 16 represents the co-occurrence of terms from the Scopus database. Figure 17 represents the co-occurrence of terms from the Web of Science database.

5. Discussion

The results of the descriptive and network analysis revealed a number of interesting insights. The search string utilised on the Scopus and Web of Science databases showed that the research on blockchain technology adoption in the area of supply chain is still in its nascent phase with a relatively limited number of publications. A total of 670 article were found between the two databases addressing the search string with the majority appearing post 2016. The descriptive analysis highlighted that

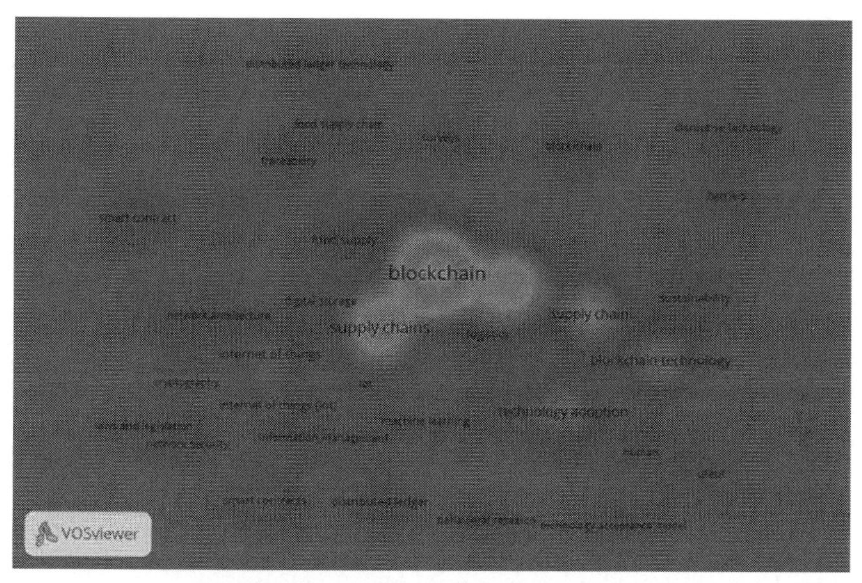

Figure 14. Co-occurrence of keywords—Scopus.

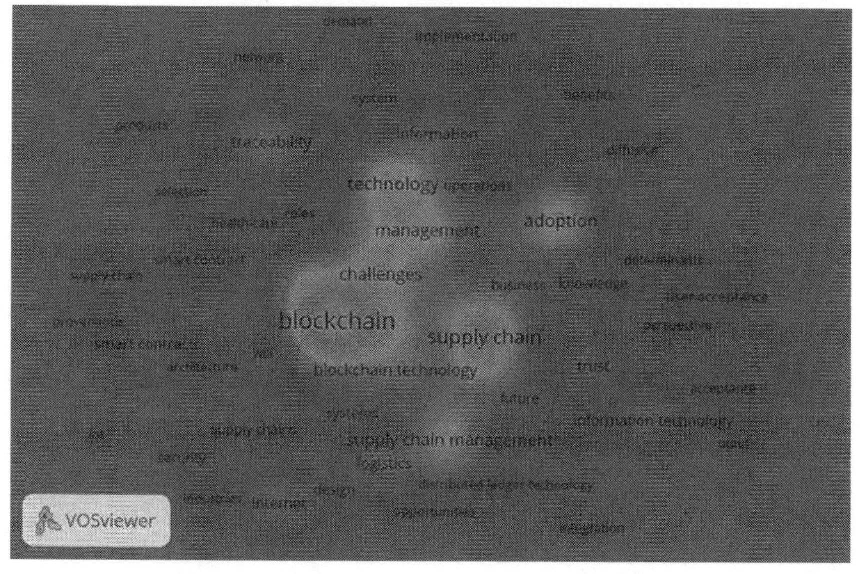

Figure 15. Co-occurrence of keywords—Web of Science.

the majority of articles fell under the auspices of production, operations, computer science, engineering, management, sustainability, and logistics. The majority of these publications are journal articles and conference papers.

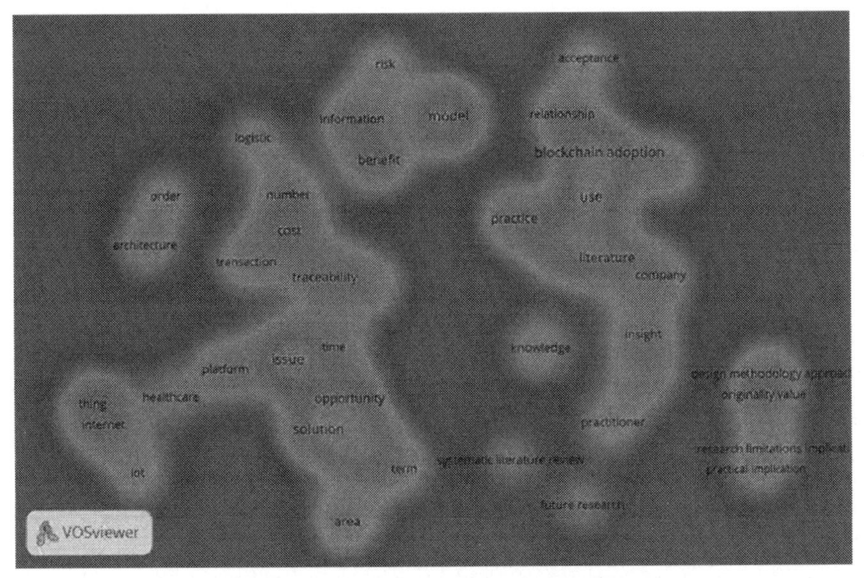

Figure 16. Co-occurrence of terms—Scopus.

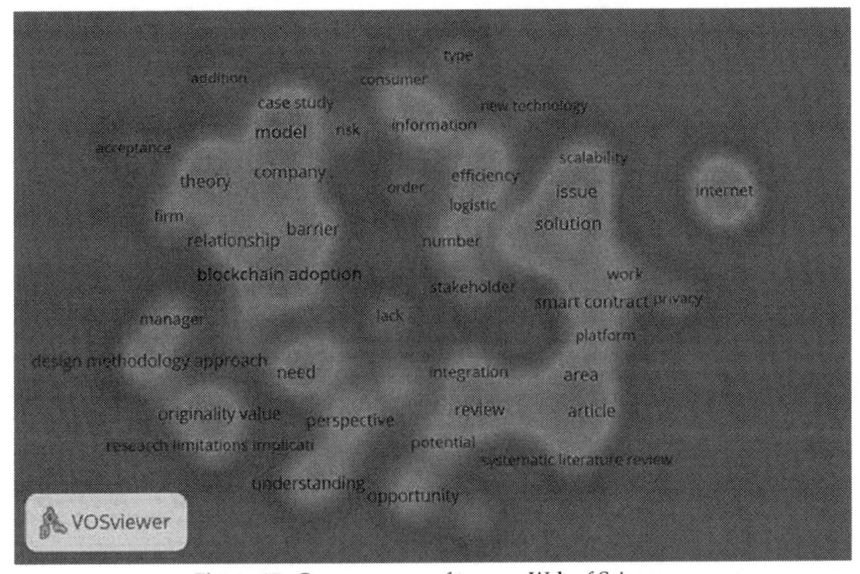

Figure 17. Co-occurrence of terms—Web of Science.

The network visualisation analysis proved a fruitful avenue as it highlighted terms and keywords that occurred with each other in the selected publications as well as the volume of occurrence. The network visualisations, Figures 6–9, measured the terms and keywords originating from the title, abstract and text of the article. The circles represent these terms and keywords with the circle size determined by the volume of occurrences, i.e., the larger the circle, the heavier the weighting. The distance between the circles denote the volume of co-occurrence of keywords/terms, i.e., the closer the circles, the higher the co-occurrence. Figure 6 shows that the keyword blockchain is closely associated with traceability and food supply. The keyword supply chain is closely associated with digital storage, internet of things, network architecture, information management and smart contracts. The keyword technology adoption is linked clearly to logistics, blockchain and supply chain. Figure 7 shows the same graphical representation as Figure 6 but is derived from Web of Science. This visualisation highlighted a much tighter clustering and can be interpreted as showing a much higher co-occurrence of keywords. The difference between the two visualisations may be explained by the differing journals hosted in each database. In this visualisation, the keyword adoption is heavily related to knowledge, trust, and acceptance. The keyword supply chain management co-occurs frequently with integration, distributed ledger technology, security, and internet of things. Finally, blockchain and smart contract are clustered tightly together. Figure 8 shows the co-occurrence of terms from Scopus. This visualisation offered a clearer delineation of occurrence and broadly fell into two areas. The first centred on the term traceability. This term co-occurred with transaction, cost, time, opportunity, benefit, and solution. The second area centred on the term blockchain adoption and was heavily related to practice, use, relationship, and acceptance. Figure 9 shows the co-occurrence of terms from Web of Science. In contrast to Scopus, the terms were more tightly clustered with a wider variety. The term blockchain adoption co-occurred frequently with relationship, barrier, need, manager, company, and model. Other terms that co-occurred frequently together were smart contract, privacy, and platform.

The overlay visualisations, Figures 10–13, are the next grouping of visualisations analysed. These visualisations represent the same data as the network visualisations but are colour coded to denote a score. This score is coded between a green and yellow with the green the lowest and yellow the highest. The score denotes the number of co-occurrence of terms or keywords. Figure 10 represents the same data as figure 6 and shows that the keywords disruptive technology, barriers, sustainability, behavioural research, and distributed ledger embody the largest volume. Figure 11 represents the same data as Figure 7 and shows that the keywords barriers, sustainability, architecture, big data, trust, perspective, and user acceptance represent the largest volume. Figure 12 represents the same data as Figure 8 and shows that the terms insight, practice, relationship, and traceability constitute the largest volume. Figure 13 represents the same data as Figure 9 and shows that there is a larger spread of terms occurring less frequently. The most frequently occurring terms are smart contract, relationship, acceptance, and barrier.

Finally, the density visualisations are examined. Similar to the overlay visualisations, colours range from blue to green to yellow. The larger the number of items in the area of a point and the higher the weights of the neighbouring items, the closer the colour of the point is to yellow. The keywords, Figures 14 and 15, emerging from both databases centred on the search string produced similar results with blockchain, technology, supply chain, and challenges emerging as the densest keywords. Figures 16 and 17 threw up broader results as the focus was on terms. Figure 16 shows dense clusters of terms in different areas of the visualisation. For example, the terms blockchain adoption, relationship, acceptance, practice, and use represent a dense co-occurrence of terms. Figure 17 from Web of Science shows less dense co-occurrence of terms but interestingly, the densest cluster is very similar to Figure 16 with the terms blockchain adoption, barrier, relationship, theory, company, and model co-occurring most often together.

6. Conclusion

This study illustrates the outcome of a descriptive and bibliometric analysis of a total of 670 articles on blockchain, supply chain and adoption emanating from two of the largest academic publishing databases, Scopus, and Web of Science. This study was conducted using Microsoft Power BI and VOSviewer. The results indicate that research at the nexus of blockchain, supply chain and adoption is still at an early stage but is steadily increasing. The majority of research is published as journal articles or conference papers. The fields/disciplines in which this research has been published most frequently are business, management, computer science, engineering, and decision science. The bibliometric analysis highlighted keywords and terms that occur frequently together in literature. The visualisations emphasised the avenues of research that are currently being pursued on these topics and also highlighted neglected research areas. We suggest further research at frequent intervals to track the research evolution of this topic. Other publications have examined blockchain in supply chain utilising bibliometric analysis. This is the first publication to examine blockchain adoption in supply chain by employing bibliometric analysis. These are the key contributions of this research. The limitations of this research are clear. As the data was extracted from two academic databases the results only cover the range and scope of journals contained within these, excluding a number of other platforms.

References

[1] Wang, Y., Han, J. H. and Beynon-Davies, P. 2019. Understanding blockchain technology for future supply chains: a systematic literature review and research agenda. *Supply Chain Management*, vol. 24, no. 1. Emerald Group Holdings Ltd., pp. 62–84, Mar. 04, 2019, doi: 10.1108/SCM-03-2018-0148.

[2] Dutta, P., Choi, T.-M., Somani, S. and Butala, R. 2020. Blockchain technology in supply chain operations: Applications, challenges and research opportunities. *Transp. Res. Part E Logist. Transp. Rev.*, 142, doi: 10.1016/j.tre.2020.102067.

[3] Yaga, D., Mell, P., Roby, N. and Scarfone, K. 2018. Blockchain technology overview.

[4] Nakamoto, S. 2008. Bitcoin: A peer-to-peer electronic cash system. *Decentralized Bus. Rev.*, p. 21260.

[5] Clohessy, T. 2019. *Blockchain: The Business Perspective.* NovoRay Publishers.

[6] Crosby, M., Pattanayak, P., Verma, S. and Kalyanaraman, V. 2016. Blockchain technology: Beyond bitcoin.*Appl. Innov.*, 2: 6–10, p. 71.

[7] Rejeb, A., Rejeb, K. and Keogh, J. G. 2021. Centralized vs. decentralized ledgers in the money supply process: a SWOT analysis. *Quant. Financ. Econ.*, 5(1): 40–66, doi: 10.3934/qfe.2021003.

[8] Swan, M. and De Filippi, P. 2017. Toward a philosophy of blockchain: A symposium: Introduction. *Metaphilosophy*, 48(5): 603–619.

[9] Abebe, E. et al. 2019. Enabling Enterprise Blockchain Interoperability with Trusted Data Transfer (industry track)', in *Middleware Industry 2019 - Proceedings of the 2019 20th International Middleware Conference Industrial Track, Part of Middleware 2019*, 2019, pp. 29–35, doi: 10.1145/3366626.3368129.

[10] Puthal, D., Malik, N., Mohanty, S. P., Kougianos, E. and Das, G. 2018. Everything you wanted to know about the blockchain: Its promise, components, processes, and problems', *IEEE Consum. Electron. Mag.*, 7(4): 6–14.

[11] Natarajan, H., Krause, S. and Gradstein, H. 2017. Distributed ledger technology and blockchain.

[12] Saberi, S., Kouhizadeh, M. and Sarkis, J. 2019. Blockchains and the Supply Chain: Findings from a Broad Study of Practitioners. *IEEE Eng. Manag. Rev.*, 47(3): 95–103, doi: 10.1109/EMR.2019.2928264.

[13] Queiroz, M. M. and Fosso Wamba, S. 2019. Blockchain adoption challenges in supply chain: An empirical investigation of the main drivers in India and the USA', *Int. J. Inf. Manage.*, 46: 70–82, Jun. 2019, doi: 10.1016/j.ijinfomgt.2018.11.021.

[14] Kumar Bhardwaj, A., Garg, A. and Gajpal, Y. 2021. Determinants of Blockchain Technology Adoption in Supply Chains by Small and Medium Enterprises (SMEs) in India. *Math. Probl. Eng.*, pp. 1–14, Jun. 2021.

[15] Allen, D. W. E., Berg, C., Davidson, S., Novak, M. and Potts, J. 2019. International policy coordination for blockchain supply chains. *Asia Pacific Policy Stud.*, 6(3): 367–380, Sep. 2019, doi: 10.1002/app5.281.

[16] Surjandy, Meyliana, H. L. H. Spits Warnars and Abdurachman, E. 2020. Blockchain Technology Open Problems and Impact to Supply Chain Management in Automotive Component Industry 2020, doi: 10.1109/ICCED51276.2020.9415836.

[17] Alazab, M., Alhyari, S., Awajan, A. and Abdallah, A. B. 2021. Blockchain technology in supply chain management: an empirical study of the factors affecting user adoption/acceptance', *Cluster Comput.*, 24(1): 83–101, Mar. 2021, doi: 10.1007/s10586-020-03200-4.

[18] Gokalp, E., Coban, S. and Gokalp, M. O. 2019. Acceptance of Blockchain Based Supply Chain Management System: Research Model Proposal', in *1st International Informatics and Software Engineering Conference, IISEC 2019,* doi: 10.1109/UBMYK48245.2019.8965502.

[19] Oliveira, T. and Martins, M. F. 2011. Literature review of information technology adoption models at firm level. *Electron. J. Inf. Syst. Eval.*, 14(1): 110–121.

[20] Tornatzky, L. G., Fleischer, M. and Chakrabarti, A. K. 1990. *Processes of technological innovation*. Lexington books.

[21] Clohessy, T., Treiblmaier, H., Acton, T. and Rogers, N. 2020. Antecedents of blockchain adoption: An integrative framework', *Strateg. Chang.*, 29(5): 501–515, Sep. 2020, doi: 10.1002/jsc.2360.

[22] Rogers, E. M. 2010. *Diffusion of innovations*. Simon and Schuster.

[23] Wang, Y.-M., Wang, Y.-S. and Yang, Y.-F. 2010. Understanding the determinants of RFID adoption in the manufacturing industry. *Technol. Forecast. Soc. Change*, 77(5): 803–815.

[24] Moosavi, J., Naeni, L. M., Fathollahi-Fard, A. M. and Fiore, U. 2021. Blockchain in supply chain management: a review, bibliometric, and network analysis', *Environ. Sci. Pollut. Res.*, 1–15.

[25] Tandon, A., P. Kaur, M. Mäntymäki and A. Dhir. 2021. Blockchain applications in management: A bibliometric analysis and literature review. *Technol. Forecast. Soc. Change*, 166: 120649.

[26] Rouzbahani, H. M., Karimipour, H., Dehghantanha, A. and Parizi, R. M. 2020. Blockchain applications in power systems: A bibliometric analysis', in *Blockchain Cybersecurity, Trust and Privacy*, Springer, pp. 129–145.

[27] Lardo, A., Corsi, K., Varma, A. and Mancini, D. 2022. Exploring blockchain in the accounting domain: a bibliometric analysis. *Accounting, Audit. Account. J.*

[28] Reis-Marques, C., Figueiredo, R. and de Castro Neto, M. 2021. Applications of Blockchain Technology to Higher Education Arena: A Bibliometric Analysis', *Eur. J. Investig. Heal. Psychol. Educ.*, 11(4): 1406–1421.

[29] Kamran, M., Khan, H. U., Nisar, W., Farooq, M. and Rehman, S.-U. 2020. Blockchain and Internet of Things: A bibliometric study. *Comput. Electr. Eng.*, 81: 106525.

[30] Niknejad, N., Ismail, W., Bahari, M., Hendradi, R. and Salleh, A. Z. 2021. Mapping the research trends on blockchain technology in food and agriculture industry: A bibliometric analysis. *Environ. Technol. Innov.*, 21: 101272.

[31] Ante, L. 2021. Smart contracts on the blockchain–A bibliometric analysis and review', *Telemat. Informatics*, 57: 101519.

[32] Müßigmann, B., von der Gracht, H. and Hartmann, E. 2020. Blockchain technology in logistics and supply chain management—A bibliometric literature review from 2016 to January 2020. *IEEE Trans. Eng. Manag.*, 67(4): 988–1007.

[33] Pandey, V., Pant, M. and Snasel, V. 2022. Blockchain technology in food supply chains: Review and bibliometric analysis. *Technol. Soc.*, p. 101954.

[34] Rejeb, A., Rejeb, K., Simske, S. and Treiblmaier, H. 2021. Blockchain technologies in logistics and supply chain management: a bibliometric review. *Logistics*, 5(4): 72.

[35] Van Eck, N. J. and Waltman, L. 2010. Software survey: VOSviewer, a computer program for bibliometric mapping. *Scientometrics*, 84(2): 523–538.

[36] Perianes-Rodriguez, A., Waltman, L. and Van Eck, N. J. 2016. Constructing bibliometric networks: A comparison between full and fractional counting. *J. Informetr.*, 10(4): 1178–1195.

[37] Mascarenhas, C., Ferreira, J. J. and Marques, C. 2018. University–industry cooperation: A systematic literature review and research agenda. *Sci. Public Policy*, 45(5): 708–718.

[38] Van Eck, N. J. and Waltman, L. 2013. VOSviewer manual', *Leiden: Univeristeit Leiden*, 1(1): 1–53.

7

Blockchain for Product Traceability in the Supply Chain

Gary Lee

1. Introduction

Bread in London was closely monitored in the late 1200s and early 1300s [1]. When bread was under-weight or rotten it was traced back to its' maker and punishments were imposed. Some of the punishments included pillory and hurdle. Pillory is the public humiliation of placing them in a wooden framework with holes for the head and hands and subjected to public abuse. Hurdle, as a punishment, was placing the baker on a sled like device and dragging the baker through the streets of town, again, for public abuse.

Tracing the underweight or bad bread to its' maker was easy to do since all commercially sold bread had to be stamped by the baker with an official seal. Today, tracing bread to its' maker is just as easy since it is typically packaged with marketing identifiers. However, it is unlikely you will see the CEO of a modern bread company in a pillory any time soon.

Modern bread makers do not place an official stamp on the bread itself. Instead, bread makers will mark the packaging with information that allows the product to be traced back to where and when it was produced. This process of tracing a product (either upstream or downstream) is referred to as product traceability. Many companies do this manually on paper or in centralized databases. This leads to slow or delayed searching

6957 N. Hwy 125, Strafford, MO 65757.
Email: glee@yahoo.com

for data with the possibility of altering the data. These and other issues exist around traditional traceability systems. Blockchain applied to the supply chain can include this traceability information and solve these issues.

2. Blockchain

Blockchain was originally designed for the cryptocurrency of bitcoin. Since the initial inception of the technology, it has been applied to many other uses. Initially the uses have been transactional. Since blockchain can be applied to any type of transaction, this opens the possibilities for many other blockchain applications. In the past few years, the supply chain has been advocated as a use case for blockchain [2].

Purchasing goods normally involves the exchange of currency. In modern supply chains this involves a third party such as a bank or broker. Satoshi Nakamoto's 2008 paper first proposed a distributed ledger eliminating the need for a trusted third-party payment system [3]. The idea allows for peer-to-peer payments using a digital currency across a public ledger system. This public ledger system is solved in blocks making it almost impossible to forge. These blocks could not be altered but only added to, which created a chain of these blocks. Thus, it was titled blockchain.

Before discussing how blockchain can be applied to product traceability we need to understand the basics of how blockchain works. Most readers of this text will be familiar with the basic representation of the blockchain where each block has the hash, timestamp, the transactional data, and other information.

In the visual representation of Figure 1 there is a section in each block for "other information". This other information is where product traceability data can be reposited.

This illustration shows the basic concepts of blockchain technology. The blockchain works well to eliminate the need for a trusted third party for ensuring transactions. The blockchain concept can also be applied to supply chain transactions.

Remember that supply chain transactions traditionally require a third party to facilitate the payments. This third party would normally be a bank. A bank would have a record of funds available. When a transaction comes through for funds to be paid to a supplier, a bank would process the transaction to the supplier. This process naturally would incur a fee that is paid to the bank for performing the transaction. It is these seemingly small transaction fees that add up to large amounts of money when hundreds, thousands, or tens of thousands of transactions are processed by companies.

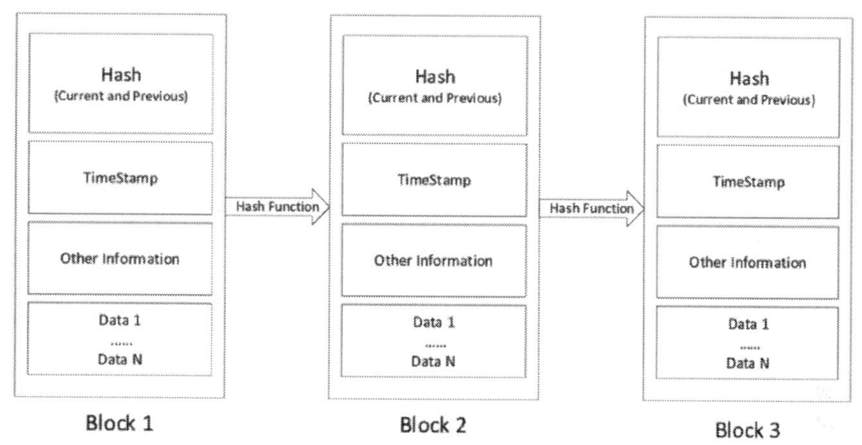

Figure 1. Basics of Blockchain.

(Reprinted with permission from Gary Lee, *Critical Success Factors for Implementing Blockchain in the Supply Chain for Product Traceability* (Doctoral dissertation, Indiana State University, 2021), 6.)

Large companies (and even small ones) can use blockchain to avoid these fees or to collect fees for themselves. This is a big incentive for companies to reduce their transaction costs. Companies like IBM, Kodak, Wal-Mart, SAP, Oracle, and Maersk have all invested millions of dollars into researching and establishing blockchains for their supply chain or their customers [4].

In addition to avoiding transactional fees as the reason to implement a blockchain in the supply chain, there are other use cases that make the implementation of distributed ledger technology a growing trend [5].

The use case of interest here is that of product traceability. Traceability is always a concern for companies. Wal-Mart and IBM have been partnered since 2016 to develop blockchain uses [6]. Frank Yiannas is Wal-Mart's Vice President of food safety. In 2017 he had his team trace the path of a package of sliced mangoes. To follow the trail using conventional recordkeeping methods from store back to farm it took 6 days, 18 hours, and 26 minutes of linear time to locate the records. Using the blockchain software they built with IBM, the same traceability exercise took 2.2 seconds [6].

Based on these anecdotal tests, it appears that Blockchain may be suitable for traceability queries even though the development initially was transaction focused. The next step then is to look deeper and determine how these systems are implemented. There are two basic formats for the implementation of a blockchain; public, and permissioned (sometimes called private).

Public blockchains are exactly as the name sounds. They are open to the public. Anyone with a way to access the network can become a node on the network. Once on the network, the node has full permissions to read, transact, and create blocks. In a permissioned blockchain, someone with a way to access the network must receive permission from the "owner" of the blockchain network to have access to the network. The network "owner" can set permissions on who can read information on the blockchain, who can transact on the blockchain, and who can write new blocks to the chain [7]. In a public blockchain the nodes existing on the chain are anonymous. In a private blockchain, all the nodes represent identifiable members [8]. In this instance it would be members of the supply chain. As many companies often buy from competitors, these kinds of privacy matters come into consideration.

Permissioned blockchains are perfectly applicable to companies who would not want all their supply chain transactions available to everyone. A supplier of mangoes to Wal-Mart would not want their competitors to see prices or payment terms, for example. By having permissioned rather than public blockchains, there are some immediate benefits, but also some immediate questions.

The benefits of a permissioned blockchain network are in the control over who can see, transact, and create on the blockchain. There is an added layer of security by requiring permission to be on the blockchain network. Other benefits include increased performance of the network since the transactions are directed to one type of transaction [9]. Additional performance improvement is gained by not having to do full proof of work but rather proof of stake.

Proof of work asks that every node on the network try to hash the block and the first that accomplishes it is given a reward in the form of bitcoin or some other cryptocurrency. Proof of stake assigns one node to solve the hash. The node is assigned by the system and other nodes can later verify the solution to reach consensus. The node that is assigned to solve the hash is incentivized to perform the work by receiving a transaction fee [10]. If this is done by the owner of the blockchain, they can not only avoid transaction fees to banks but collect fees for themselves.

Along with these benefits come some questions about the security of fewer nodes and reducing the proof of work requirement. The security question can be debated on the technical side, but it is argued that it is more secure because not only is the hash computation still completed and verified, but it is done by only those with permissioned access to the blockchain [9].

Another question is how is traceability on the blockchain better than current methods? In other words, why should a company start using traceability on the blockchain when traceability systems already exist? This debate is not a part of this study. It is noted that companies are moving to blockchain to eliminate transactional costs and taking the opportunity to include traceability data. The question evaluated here is how well blockchain handles the specific requirements of product traceability.

3. Traceability Defined

The simplest definition of traceability is knowing where a product came from. Of course, it's a little more complicated than this simple definition. Dr. Joseph Juran was an important figure in the development of modern quality and defined traceability as the "ability to trace the history, application, or location of an entity by means of recorded identifications" [11]. He simplified many concepts without losing the precision of the statement. There are, however, missing elements of this definition. For instance, while you can know the history of a product (or entity as he refers to it) this definition doesn't infer that you also know the history of the components and materials that make up that final product. Another component that isn't clear from this definition is knowing where a product went after production.

For example, if there is a bad batch of bread, there is a need to know where the yeast and the flower came from. This is needed to identify if there are other lots of bread produced from a bad batch of yeast. The manufacturer would also need to know which stores bought the suspect bread to pull it from the shelves (if needed). The history of the product at various steps of production as well as the history of the product through distribution must both be included in the definition.

ISO 8402:1994 is titled as *Quality management and quality assurance—Vocabulary* and defines traceability as, "The ability to trace the history, application or location of an entity by means of recorded identifications" [12]. This is an older standard that has since been withdrawn and was replaced with ISO 9000:2000 which has undergone further revisions. The most recent edition (2015) defines traceability a little more specifically as "the ability to trace the history, application, use and location of an item or its characteristics through recorded identification data" [13].

GS1 is a not-for-profit that creates standards for business communication. The bar code system is one of their standards. GS1 has created a traceability standard. The GS1 standard is an attempt to standardize the minimum requirements for traceability systems [14]. GS1 Global Traceability Standard uses the ISO 9000:2000 definition but expands it as, "The ability to track forward the movement through specified stage(s) of the extended supply chain and trace backward the history, application or location of that which is under consideration" [14].

The GS1 definition does a good job of what Bechine, Cimino, Marcelloni and Tomasi clarified in their definition of breaking traceability into track and trace [15]. Tracking is following a product downstream in the supply chain and tracing is following a product upstream in the supply chain.

While there is still debate about the exact definition of the term and what data should be included, the concept is well understood [16,17]. Both the history of the product throughout the supply chain as well as the history of the product through distribution must be included in the definition.

Another aspect of traceability data is that it includes production information. For example, the bread manufacturer will likely have data for each batch of bread. This data will include cooking temperature and time, the specific oven used, the date it was baked, etc. The types of data that are retained are sometimes regulatory (the company is required to collect and keep) and other times it is optional. Many authors have rightly pointed out that traceability data is not just within a single company but extends to the entire supply chain [18,19,20,21]. This means that not only is the tracking and tracing data important for the supply chain, but also the data about production is important at each step in the supply chain.

4. Elements of Traceability

Within the context of track and trace, the elements of traceability can be broken down even further. Caplan provides five necessary elements of an effective traceability system [18].

1. Lot integrity control: Lot and part identification. This is data that manufacturers retain on the history of production. When an item is produced, on which machine, which shift, etc.
2. Processing control: Unique identification of each item or group of items (lot) and process data. This information is more detailed data about production. It goes beyond when an item was produced and includes data about the processing itself. For example, a loaf of bread being baked in an oven could have data retained about the oven temperature and the time the item was in the oven. Any other processing data would fit into this category.
3. Build control: data showing which items were combined to make the product and the process data when it was built at each step of the process. Data here would include things such as which lot of yeast was combined with which lot of flour to make which lot of bread.
4. Inspection and test: records of test, rework, and other off standard work on a product. If a product has a test required or performed, then that data would be of this category. A loaf of bread might require that some of the flour is tested for coarseness during processing. The results of these kinds of tests and inspections would be retained here.

It can serve to satisfy a regulatory compliance obligation, or it could be a simple historical record if there is ever a need for this data in the future.

5. Field activity and modification control: records of field installation, service, post-delivery changes, etc. This data is added to existing records after manufacture. If a complex or safety critical piece of equipment is changed out, this information becomes important. This is highly important in the aerospace industry. For example, records that the person who installed a critical component on an aircraft was a qualified technician is important. Records that they properly tested the installation, etc., are also critical to retain.

This data can be gathered and stored in many ways. Bar codes, Radio Frequency Identification (RFID) tags, and human readable are some of the more common ways of gathering the data [22]. The ways to gather this data is beyond the scope of this review. How the data is stored is what is being evaluated. These 5 elements of traceability are excellent use cases for blockchain. If the records are gathered electronically then they can easily be transferred and stored on the supply chain blockchain. If the data is recorded on hand-written forms, then the transfer is still possible by scanning the document before transferring the electronic record to the blockchain.

5. Applications of Traceability Data

Having identified that traceability information is needed throughout the supply chain and that there are different elements of traceability data needed, the next question is when should the information be gathered and maintained? Traceability data need to be gathered and maintained when products are not identical such as different production lots or dates [16]. Töyrylä has identified 4 applications of traceability data.

1. Material flow management applications—where physically is a product. This applies primarily to shipping/logistics companies. It is not uncommon for consumers to want to track a package until it is delivered. This also applies to knowing product specific information. If a bread company is receiving a batch of yeast, what temperature was the yeast held at during shipment? This may require sensors and data collection devices but is an important source of data for food manufacturers.
2. Legal verification applications—Warranty, fraud, proof of origin, and proof of quality fall within the legal applications for traceability. These applications are necessary for the protection of both consumers and producers. When a consumer claims that a loaf of bread is moldy,

knowing that the bread was made 3 years prior is an important piece of information for the manufacturer to have. Conversely, knowing if a product is still under warranty is important for the consumer.

3. Segregation applications—used to determine which customer ordered which items. A simple example of this application is serial number tracking. In the automotive industry, for example, they use a Vehicle Identification Number (VIN) to know which consumers purchased which vehicle. This becomes important if that vehicle is involved in a recall of the product.
4. Measurement and analysis applications—used to gather data for analysis into marketing efforts, quality relationships to design changes, etc. Demographic information such as which stores are selling more or less of a product, for example, falls into this category. Studying this data gives marketers the ability to target their efforts, predict future selling trends, etc.

Based on Töyrylä applications it can easily be seen that traceability data is not limited to knowing where a product came from or went in the supply chain [16]. Other uses for traceability data that [16] points out are in logistics, quality, security, accounting, and after-sales applications. None of these applications is not beyond the capabilities of blockchain.

Many industries use traceability data [17,21]. Because of the safety requirements of the food industry, most of the literature about traceability focuses on food traceability [23,24,25,26,27]. Regardless of the industry, traceability has an important role in modern society, and blockchain can fulfill all these needs and requirements.

6. Pillars of Traceability Systems

A traceability system is a necessity, but to have an effective traceability system there are four pillars [28]. It can be shown that blockchains can satisfy each of these pillars.

Product identification is the first of the four pillars identified by Regattieri, Gamberi, and Manzini. Product identification is the fundamental information about a product (weight, volume, dimensions, perishability, cost, etc.). One of the complications of this data is that it includes the bill of material of a product. Most products are not a single item but are made up of many other components. The data here can become large but are easily stored in the blockchain at each step of the supply chain. Once it is part of the blockchain it becomes immutable. This is important for many of the applications discussed earlier.

Data to trace is the second pillar. This is essentially the characteristics of the data in the system. For instance, the kinds of data stored. This includes whether the data is numeric or alpha numeric. What are the ranges of the data? Is the data confidential? How many characters are allowed, etc., are all considerations of an effective traceability system. Again, blockchain can (and must) include data to trace as part of a traceability system.

Product routing is the third pillar defined by Regatieri et al. [28]. This data is perfectly suited for a blockchain application. How a product is routed through the supply chain and the accumulated data along the production or delivery process is what is included in this pillar. The data here can be gathered in many ways. Much of the literature around traditional traceability systems discusses improving the current systems through technological means. The most common means of improving current traceability is using RFID [23,24,25,27]. One key point with the improvement systems discussed is the assumption of a central database. Blockchain is a distributed ledger but is still able to retain these improvements.

The last pillar that is discussed is traceability tools. The fundamental component of this pillar is the accuracy and reliability. The immutability of blockchains is of primary importance here. Because the ledger is distributed and not easily changed, the reliability can be trusted. The accuracy of the data is not any different from traditional traceability systems, except that it cannot be changed after it is on the blockchain. Cost is also discussed as a consideration for this pillar. Since the implementation of blockchains in the supply chain are for the purposes of saving costs through disintermediation, the addition of traceability data would not outweigh these savings.

Blockchain can address the pillars of a traceability system. The pillars are solidly planted. How well blockchain handles the issues normally associated with traditional traceability systems is the next question to be answered.

7. Traceability Issues

There is a lot of discussion around traditional traceability systems on how to improve them. Methods of improving current legacy systems include bar codes and Radio Frequency Identification (RFID) tags [22]. These methods do a good job of improving the current systems. However, these improvements do not fundamentally change the processes. Improvements are needed where the system needs exceed the current processes. There are five common types of issues that are experienced in traceability systems. The issues experience in traceability systems fit into the following categories:

1. Real time information [29]—Being able to retrieve information in a timely manner. Having information available too late can not only

be a frustration for companies and consumers, but it can also be a legal requirement. In many industries, there are time requirements if there were to be a quality issue. For example, in the United States the National Highway Transportation Safety Administration requires that from the time an automotive manufacturer makes a recall determination the consumer must be notified by mail within 60 days [30]. Consider that the manufacturer must determine which vehicles are suspect, track which dealer received the effected vehicles, then get the records from each dealer on which consumer purchased each vehicle, and then locate any licensing records indicating the vehicle was re-sold. Each of these steps is fraught with complications in a manual process.

In a decentralized database like blockchain this process becomes much faster. Recall the example of Frank Yiannas of Wal-Mart in 2017. What took 6 days, 18 hours, and 26 minutes for a single package of Mangoes, he was able to trace in 2.2 seconds on the blockchain software they built with IBM [6].

2. Easy availability [31]—Being able to have access to the information. Using the example of a vehicle recall above, it is easy to see that this is not an insignificant consideration. If the recall is precipitated by a component manufacturer, they would not even be able to complete the first step of determining which vehicles received the suspect component. The component manufacturer would be reliant upon the vehicle manufacturer to provide any needed information. As supply chains get longer this gets exacerbated. Imagine a third-tier supplier having to initiate a recall. The process would quickly become cumbersome. With a blockchain established on the supply chain that has the traceability information included, the information becomes easily available all along the supply chain.
3. Long term storage [22]—Retaining information for extended periods of time. Staying with the automotive industry example, record retention can become very protracted. Vehicles have expected useful life well beyond a decade in many cases. How long to retain documents is not specified by quality standards like IATF 16949 but does require the manufacturer to have a policy [32]. It is not uncommon to specify this length of time as the life of the product plus some designated number of years. Traditional filing systems (i.e., paper) creates multiple difficulties including physical storage space, degradation of materials, and reliance on only one copy.

   The physical storage space of paper documents is a big consideration. In the early 2000's it was common to hire companies to mass scan documents into electronic filing systems. Having scanned copies of documents alleviates the physical storage and degradation issues but does not always solve the reliance on a single copy. Storing this data on the blockchain alleviates the physical storage space, degradation of materials, and reliance on a single copy issue. Being electronic there is no degradation of the paper and no concern with where to store the information. By being on a distributed ledger, there is multiple copies of the data if one were to become corrupted.

4. Security [16]—Preventing the data from being compromised. Like long term storage concerns, there is a concern for keeping the traceability data safe. When this data is retained in paper storage, systems must be put into place to prevent their loss. Additionally, they must be protected from damage in the event of a fire. Having duplicate copies retained in different locations is one solution, but not practical with paper-based systems. An electronic copy solves this issue by scanning to a centralized database. With centralized database systems a separate copy is often updated on a routine basis. While this solves the long-term security, it has an inherent gap between updates to the copy. Systems like RAID striping solve many of the potential data corruption issues, but do not protect against natural disasters. A more elegant solution is blockchain that is inherently secure from loss or damage and is secure from data corruption due to the distributed redundant nature of the blockchain.
5. Accuracy [16]—The data does not contain errors. Automatic or automated systems to enter traceability information is far superior to human entry. Sharp estimates that data entry even among skilled typists contains one mistake out of every 300 characters [33]. Conversely, automatic, or automated systems have approximately one mistake out of every 1 million characters. The inclusion of errors upon entry is irrelevant to the comparison of traceability systems that are centralized or stored on a blockchain. The key difference in a distributed system like blockchain is the immutability. Short of a 51% attack it is impossible to change an entry after it has been codified as part of the blockchain. This can be an advantage that supports correct records reporting (preventing falsification). It can be a disadvantage if there is human entry that resulted in errors.

Blockchain is well suited to handle the issues associated with traditional traceability systems. Improving upon traditional systems by overcoming the issues is a powerful support of the use case of blockchain for product traceability in the supply chain. Additionally, some authors have written about the potential benefits of implementing blockchain in the supply chain [19]. Tian gives both advantages of blockchain as well as disadvantages [25]. Tian does this in the context of combining blockchain technology with RFID systems. The advantages enumerated are; better tracking and tracing, enhanced credibility of safety information, and fighting against fake products. The disadvantages listed are simply the high cost (of the RFID for every product), and the immaturity of blockchain. The example given is the number of transactions per second that can be handled through blockchain. Blockchain can perform up to 7 transactions per second compared to 47,000 transactions per second that Visa processes [25]. It should be noted that since the writing of this article in 2016, the number of transactions per second on blockchain has been developed further. By 2018 (just 2 years later) Hyperledger Fabric, for example, can handle more than 3,500 transactions per second and it continues to develop [34].

Rapalis and Hossain also wrote about the potential benefits of blockchain for product traceability in the supply chain [35]. They enumerate much of what Tian list. What they also provide are some of the potential challenges to implementing blockchain in the supply chain for product traceability that they cite from the literature.

1. Lack of standardized format for information exchange in the supply chain
2. Differences in accuracy levels of traceability between links in the supply chain
3. Lack of integration and transparency within the supply chain
4. Data issues such as trust, privacy, security, and reliability

Lack of standardized format exists in traditional traceability systems of both paper and centralized databases as well. This challenge exists independent of blockchain. GS1, mentioned earlier, is working to standardize the minimum requirements for traceability systems by publishing a Global Traceability Standard [14].

Differences in accuracy between links in the supply chain is also independent of blockchain and exists in traditional traceability systems. The Global Traceability Standard from GS1 addresses this concern as well [14].

A lack of integration and transparency within the supply chain is exclusive to traditional systems. Once large companies (Wal-Mart, Maersk, IBM, etc.) begin requiring suppliers to use private blockchains to be a registered supplier, this concern becomes moot. The blockchain by its nature is integrated and transparent.

The last issue mentioned by Rapalis and Hossain is about trust, privacy, security, and reliability [35]. Biggs, Hinish, Natale and Patronick in contrast, assert that blockchain technology used in the supply chain will build trust and transparency. They see what Rapalis and Hossain call data issues as enablers and not potential challenges. They list their own challenges under the heading "Blockchain Barriers to Marketplace Acceptance" [36] which include:

1. Uncertain government regulatory status
2. Large energy consumption
3. Cross industry integration
4. Black market

These challenges are seen to be barriers to acceptance and not barriers to implementation. Malyavkina, Savina, and Parshutina begins to get more specific about challenges to implementation [37]. They list the challenges in categories of technology, organizational, normative, legal, economic, and psychological. Clohessy groups these into the categories of Technological, Organizational, and Environmental (TOE) [2]. He then divides each category out into factors affecting implementation of blockchain in the supply chain.

Lee expands upon Clohessy's critical success factors by providing a model of which are the most important to implementing blockchain in the supply chain for product traceability [38]. He found that permissions (public vs. private), organizational size, and blockchain knowledge all have a positive impact upon implementing blockchain in the supply chain.

From all of this research, it can be seen that the challenges are being overcome. Evidence of this is the progression of blockchain technology along the Gartner Hype Cycle. To evaluate a technology from the initial innovation to full acceptance the Gartner Hype Cycle proves useful. It is a graphic representation of technologies life cycle from inception to maturity. On the X axis is time and on the Y axis is expectations of the technology. Depending on where a technology is in maturity, the amount of expectation over time raises and lowers creating a wave form graph. According to Gartner Research, different parts of blockchain technologies are at different places along the graph. What Gartner terms 'authenticated provenance' can also be termed product traceability. Gartner places authenticated provenance as being on the rise in 2020 in terms of expectations of the technology [39].

As technologies continue to mature the Gartner cycle predicts a 'trough of disillusionment' as implementations fail. Preventing failed implementations are the encouraging connections to social issues.

One of the criticisms of blockchain in general is the power requirements. Traceability, on the other hand, is being researched in connection with current social issues. There has been research on the role that traceability plays in sustainability [40,41,42,43]. A measure is now in place that is used by the textile and clothing industry, called the Higg Index, that scores suppliers on traceability and sustainability [44]. Traceability as a contributor to social issues can complement the negatively perceived power requirements of blockchain.

8. Conclusion

Bread makers no longer live in fear of pillory or hurdle for underweight loaves. However, the loaves still need to meet regulatory standards, and having a traceability system is necessary for this purpose.

Initially blockchain technology was used for Bitcoin and other cryptocurrencies. In recent years organizations have explored other use cases for the technology. Blockchain is a new tool that holds a lot of promise as a developing field. Traceability is a new use case for this technology and companies are investing heavily in development.

Investment is ongoing as companies work to develop blockchain technology. One of the main goals of the technology is disintermediation. By removing the banks and brokers from the payment process in the supply chain, transactional costs can be eliminated. By eliminating the supply chain transactional costs, the benefits can be monetarily beneficial. Use cases of blockchain are being explored. One of the identified advantages of blockchain beyond disintermediation is to incorporate traceability. This immutable database of provenance makes traceability a natural fit with blockchain technology. This research looked at the use of blockchain technology in supply chain traceability.

After defining traceability to include track and trace, it was shown how there is a natural fit to use blockchain in the supply chain. Disintermediation is the impetus for supply chain transactions. Another necessary thing in supply chain transactions is product traceability. Since these traceability transactions can be included in the blocks of a blockchain, it becomes a natural fit.

Even though it is a fit within the structure of the blockchain, a traceability system must satisfy the five fundamental elements of a traceability system (Lot integrity control, Processing control, Build control, Inspection and test, Field activity and modification control).

It was shown that not only does blockchain satisfy the fundamental elements of a traceability system, but it also satisfies the four pillars of a traceability system (Product identification, Data to trace, Product routing, Traceability tools).

From this it is easy to believe that using blockchain in the supply chain can satisfy the traceability requirements. Beyond being able to satisfy the fundamental elements of a traceability system and the four pillars of a functional traceability system, blockchain also eliminates many of the issues that plague traditional traceability systems (real time information, availability, long term storage, security, accuracy).

Once the issues of traditional traceability systems were shown to be overcome by blockchain traceability systems, there comes the issue of implementing blockchain in the supply chain. Implementing blockchain has many issues. Four were identified by Rapalis and Hossain as; Lack of standardized format, Differences in accuracy, Lack of integration and transparency, Data issues such as trust, privacy, security, and reliability [35].

While Biggs, Hinish, Natale, and Patronick have other implementation challenges that include uncertain government regulatory status, large energy consumption, cross industry integration, black markets [36].

All these issues are overcome in different ways. Some are issues that exist within the supply chain regardless of the use of blockchain, while others are uncertainty or perception issues. The perception issues are of public blockchain rather than the private blockchains that are to be used within supply chains.

This thorough look at product traceability in the supply chain shows that blockchain can solve many, if not all, of the issues with traditional systems. The use case of blockchain in the supply chain is feasible. The technology is a natural fit for traceability systems across multiple entities. Additionally, traceability has a positive social impact. As social issues become more important to consumers, measures like the Higgs index may play a more important role in consumer choices that the marriage of traceability with blockchain in the supply chain can satisfy. Enjoy the next loaf of bread being assured that blockchain can support the supply chain in maintaining product traceability.

Glossary

Product Traceability—Using recorded data to know how a product was produced, where it was produced, where the constituent components came from, where the product went after it was produced, and processing information about the product.

Product Provenance—Similar to product traceability and often used interchangeably. Provenance is only looking in one direction in the supply chain to determine a product's origin.

Gartner Hype Cycle—Gartner is a research and advisory company that produced graphic representation of technologies life cycle from inception to maturity.

Higgs index—A standardized measurement of value chain sustainability using 5 core tools.

Track—Following a product's data trail downstream in the supply chain.

Trace—Following a product's data trail upstream in the supply chain.

RAID striping—Redundant Array of Independent Disks is a method of data storage that stores multiple copies of data across multiple disks in smaller parts so that if any one disk is compromised, the entire data set can be rebuilt.

Disintermediation—Reducing or eliminating intermediaries. In the supply chain it would be paying a supplier directly instead of using a bank (the intermediary) to perform the transaction.

RFID tags—Radio Frequency IDentification tags are small radio transmitters that transmit data about a product to a reader. RFID tags can be read from a short distance rather than having to get closer to scan a bar code or read a tag.

References

[1] Cosman, M. P. 1976. *Fabulous feasts: Medieval cookery and ceremony.*

[2] Clohessy, T. 2019. *Blockchain The Business Perspective* (1st ed.). Galway, Ireland: NovoRay.

[3] Nakamoto, S. 2008. *Bitcoin: A peer-to-peer electronic cash system.* (Unpublished white paper). Retrieved from https://bitcoin.org/bitcoin.pdf

[4] Bowles, J. 2018. How IBM's blockchain investments are paying off. Retrieved from https://diginomica.com/2018/01/12/how-ibms-blockchain-investments-are-paying-off/

[5] Yli-Huumo, J., Ko, D., Choi, S., Park, S. and Smolander, K. (2016). Where is current research on blockchain technology?—a systematic review. *PloS one*, 11(10): e0163477.

[6] McKenzie, J. 2018. Why blockchain won't fix food safety—yet. Retrieved from https://newfoodeconomy.org/blockchain-food-traceability-walmart-ibm/

[7] Bauerle, N. (n.d.). *What is the Difference Between Public and Permissioned Blockchains?* Retrieved from https://www.coindesk.com/information/what-is-the-difference-between-open-and-permissioned-blockchains/.

[8] Pilkington, M. 2016. 11 Blockchain technology: principles and applications. *Research handbook on digital transformations*, 225.

[9] Monax. (n.d.). *Learn / Permissioned Blockchains.* Retrieved from https://monax.io/learn/permissioned_blockchains/.

[10] Blockgeeks. 2017. *Proof of Work vs Proof of Stake: Basic Mining Guide.* Retrieved from https://blockgeeks.com/guides/proof-of-work-vs-proof-of-stake/.

[11] Juran, J. and Godfrey, A.B. 1999. *Juran's quality handbook: The complete guide to performance excellence.* McGraw-Hill Education.

[12] International Organization for Standardization. 1994. *ISO 8402: 1994: Quality Management and Quality Assurance-Vocabulary.* International Organization for Standardization.

[13] International Organization for Standardization. 2015. *ISO 9001: 2015: Quality Management Systems – Requirements.* International Organization for Standardization.

[14] Ryu, J. and Taillard, D. 2007. GS1 Global Traceability Standard: Business Process and System Requirements for Full Chain Traceability. *GS1 Standards Document. GS1.*

[15] Bechini, A., Cimino, M., Marcelloni, F. and Tomasi, A. 2008. Patterns and technologies for enabling supply chain traceability through collaborative e-business. *Information and Software Technology*, 50(4): 342–359, doi: 10.1016/j.infsof.2007.02.017.

[16] Töyrylä, I. 1999. *Realising the potential of traceability: a case study research on usage and impacts of product traceability.* Helsinki University of Technology.

[17] Hobbs, J. E. 2003. Consumer demand for traceability. *Retrieved January*, 28, 2006.

[18] Caplan, F. 1989. The Quality System: A Sourcebook for Managers and Engineers, Chilton Book Company, Pennsylvania.

[19] Abeyratne, S. A. and Monfared, R. P. 2016. Blockchain ready manufacturing supply chain using distributed ledger.

[20] Kim, H. M. and Laskowski, M. 2018. Toward an ontology-driven blockchain design for supply-chain provenance. *Intelligent Systems in Accounting, Finance and Management, 25*(1): 18–27.

[21] Espinoza Limón, A. and Garbajosa Sopeña, J. 2005. The need for a unifying traceability scheme.

[22] Steele, D. C. 1995. A structure for lot-tracing design. *Production and Inventory Management Journal*, 36(1): 53.

[23] Opara, L. U. 2003. Traceability in agriculture and food supply chain: a review of basic concepts, technological implications, and future prospects. *Journal of Food Agriculture and Environment*, 1: 101–106.

[24] Dabbene, F., Gay, P. and Tortia, C. 2014. Traceability issues in food supply chain management: A review. *Biosystems Engineering*, 120: 65–80.

[25] Tian, F. (2016, June). An agri-food supply chain traceability system for China based on RFID & blockchain technology. In *Service Systems and Service Management (ICSSSM), 2016 13th International Conference on* (pp. 1–6). IEEE.

[26] Aung, M. M. and Chang, Y. S. 2014. Traceability in a food supply chain: Safety and quality perspectives. *Food control*, 39: 172–184.

[27] Kelepouris, T., Pramatari, K. and Doukidis, G. 2007. RFID-enabled traceability in the food supply chain. *Industrial Management & Data Systems*, 107(2): 183–200.

[28] Regattieri, A., Gamberi, M. and Manzini, R. 2007. Traceability of food products: General framework and experimental evidence. *Journal of Food Engineering*, 81(2): 347–356.

[29] Feigenbaum, A. 1991. Total quality control, McGraw-Hill, Inc., 3rd ed., Singapore.

[30] NHTSA. (n.d.). *Roles in the Recall Process*. Retrieved from https://www.nhtsa.gov/recalls

[31] Martin, J. 1983. *Managing the data base environment*. Prentice Hall PTR.

[32] IATF 16949. 2016 *Quality Management System for organizations in the automotive industry*

[33] Sharp, K. R. 1990. *Automatic Identification: making it pay*. Van Nostrand Reinhold Computer. Shrier, D., Sharma, D. and Pentland, A. 2016. Blockchain & financial services: The fifth horizon of networked innovation. *White paper excerpt, MIT*.

[34] Androulaki, E., Barger, A., Bortnikov, V., Cachin, C., Christidis, K., De Caro, A., Enyeart, D., Ferris, C., Laventman, G., Manevich, Y., Murthy, C., Nguyen, B., Sethi, M., Singh, G., Smith, K., Sorniotti, A., Stathakopoulou, C., Vukolic, Cocco, S., Yellick, J. and Muralidharan, S. (2018, April). Hyperledger fabric: a distributed operating system for permissioned blockchains. In *Proceedings of the Thirteenth EuroSys Conference* (p. 30). ACM.

[35] Rapalis, G. and Hossain, S. 2019. Traceability in the Food Industry.

[36] Biggs, J., Hinish, S. R., Natale, M. A. and Patronick, M. 2017. Blockchain: Revolutionizing the Global Supply Chain by Building Trust and Transparency. *Rutgers University, New Jersey*.

[37] Malyavkina, L. I., Savina, A. G. and Parshutina, I. G. (2019, May). Blockchain technology as the basis for digital transformation of the supply chain management system: benefits and implementation challenges. In *1st International Scientific Conference" Modern Management Trends and the Digital Economy: from Regional Development to Global Economic Growth"(MTDE 2019)*. Atlantis Press.

[38] Lee, G. 2021. *Critical Success Factors for Implementing Blockchain in the Supply Chain for Product Traceability* (Doctoral dissertation, Indiana State University).

[39] Litan, A. and Leow, A. 2020. Hype Cycle for Blockchain Technologies. Retrieved from https://www.gartner.com/en/documents/3987450/hype-cycle-for-blockchain-technologies-2020.

[40] Germani, M., Mandolini, M., Marconi, M., Marilungo, E. and Papetti, A. 2015. A system to increase the sustainability and traceability of supply chains. *Procedia CIRP*, 29: 227–232.

[41] Busse, C., Meinlschmidt, J. and Foerstl, K. 2017. Managing information processing needs in global supply chains: A prerequisite to sustainable supply chain management. *Journal of Supply Chain Management*, 53(1): 87–113.

[42] Aarseth, W., Ahola, T., Aaltonen, K., Økland, A. and Andersen, B. 2017. Project sustainability strategies: A systematic literature review. *International Journal of Project Management*, 35(6): 1071–1083.

[43] Badzar, A. 2016. Blockchain for securing sustainable transport contracts and supply chain transparency-an explorative study of blockchain technology in logistics.

[44] Agrawal, T. K. 2019. *Contribution to development of a secured traceability system for textile and clothing supply chain* (Doctoral dissertation, Högskolan i Borås).

8

Benefits and Barriers to Blockchain Adoption in Industry 4.0 and the Circular Economy

Garry Lohan

1. Introduction

There is little doubt that emerging technologies in supply chains, across industries, will continue to help organisations better fulfil their customers' needs and improve the overall traceability and accountability of their upstream suppliers. Blockchain Technology (BCT) in particular, is touted as a potential solution to many of the issues currently faced by organisations where traceability and accountability is paramount. BCT not only promises to be a solution to external upstream supply issues but also for similar issues faced by internal value supply chains. Blockchain is a distributed ledger technology whereby transparency and immutability are inbuilt into the supply chain process. Once a product or service is entered into the chain the actors in the process can be sure the data is trustworthy. This issue with trust in supply chain is of utmost importance to most industries and of critical importance to several industries such as healthcare, military and the food industry. Implementing blockchain into supply chains is a decision that individual actors cannot make in isolation. There must be a minimum of two actors to create any block in the chain and usually transforming legacy supply chain processes involves a

Atlantic Technological University, Dublin Road, Galway, Ireland.
Email: garry.lohan@gmit.ie

strategic internal organisational and cross-organisational decision as the use of blockchain involves a change in process and workflow that is often met with organisational resistance to the detriment of the overall strategy.

Efforts to streamline any new implementation process can be aided by theories that outline behavioural concerns. As with any new technology, acceptance criteria are guided by the end-goal of getting actors to use the new technology. New technology acceptance models such as the technology acceptance model (TAM), the resource-based view of the firm (RBV) or the Business Model may be of use to implementation managers who can use lessons learned from these theories to help plan and implement the transformation. While there are alternatives and BCT is not always the answer, it does present promising opportunities for improvement across the supply chain. Using blockchain in the supply chain has the potential to improve supply chain transparency and traceability as well as reduce administrative costs and as Industry 4.0 (I4.0) ramps up automation and streamlining of processes BTC is well positioned to become a major mechanism through which organisations can economically achieve their specified objectives within the framework of what is often termed the new Circular Economy of the 21st century.

2. Technology Transformation Acceptance

Industry 4.0 adoption in supply chain management for the circular economy implies a disruptive change in the business model. Implementing technologies such as BCT requires a change management program for most organisations and the change management literature has long highlighted the issues with new processes and ways of working. Socio-economic challenges in adopting technological innovations are also well documented. For example, [1] suggest the adoption of I4.0 in supply chain is a digital transformation process that may take years to get accepted by stakeholders. [2] argue that the product, process, market, and environment all present major barriers that need to be overcome. There are also other challenges faced with implementing a blockchain database. Organisational challenges revolving around legal and regulatory frameworks, environmental challenges around energy consumption, social challenges such as a lack of required technical skill or even cultural challenges in moving from a centralised to decentralised network which will require the buy-in of its users and operators. These, coupled with the technological challenges of integrating, scaling, replacing or complementing legacy systems, the availability of supporting infrastructure, the cyber-security concerns of control, security, and privacy and throughput and latency issues which currently limit the potential to conduct large data analytics in real time—show that blockchain adoption is still in its early days and it is useful to further understand how this new technology is being used and whether its potential is being fully realised.

Many theoretical perspectives have been developed to help understand how users of a new technology make decisions to use this new technology. These theories or frameworks can provide tools and guidance to understand success or failure in implementation processes of new IT applications. As with any theory, they have evolved over time and as newer technologies become mainstream, the theories are continuously expanded and adapted as and when or if required. In this section, we look at some examples of theories which may prove useful in understanding the Blockchain environment. This is by no means an overarching list of frameworks available, but it does provide us with some guidance when considering a blockchain implementation into a supply chain. The theories proposed are the:

- Technology Acceptance Model (TAM) [3]
- Dynamic Capabilities Model [4]
- Business Model [5]
- Unified Theory of Acceptance and Use of Technology (UTAUT) [6]

Table 1. Theoretical Frameworks.

Theory	**Main Independent Construct(s)**	**Main Dependent Construct(s)**
TAM	Perceived usefulness Perceived ease of use	Behavioural intention to use System usage
Dynamic Capabilities	Capabilities Absorptive capacity Environmental turbulence Agility	Sustainable competitive advantage
The Business Model	Value proposition Target customer segment Core competency The competitive positioning The economic model The investment model	Decision making levels
Unified Theory of Acceptance and Use of Technology	Gender Age, Voluntariness Experience	Performance expectancy Effort expectancy Social influence Facilitating conditions
The Resource-Based View of the Firm	Assets, capabilities, resources	Competitive advantage Organizational performance
Diffusion of Innovations Theory	Compatibility of technology Complexity of technology Relative advantage	Implementation success Technology adoption

- Resource-Based View of the Firm [7]
- Diffusion of Innovations Theory (DOI) [8]

The Technology Acceptance Model (TAM) in particular, has received widespread use within academia and industry and is the most frequently used among all other theories [9]. Results from TAM studies are not always consistent however and for this reason, plus the fact that end user acceptance issues may not be the main barrier to blockchain adoption, the following theories are positioned as complementary to TAM for studying stages of the transformation process.

2.1 The Technology Acceptance Model

TAM theory is based on principles adopted from [10] in psychology, which show how to measure the behaviour-relevant components of attitudes and specify how external stimuli are causally linked to beliefs, attitudes and behaviour. The theoretical model on which TAM is based is the Theory of Reasoned Action (TRA), a general model which is concerned with individuals' intended behaviours. According to TRA an individual's performance is determined by the individual's attitude and subjective norms concerning the behaviour in question. In addition, an individual's beliefs and motivation interact with existing behaviour. With the information technology (IT) age of the 1990s, TRA proved too general, and TAM was developed specifically to predict individual adoption and use of new ITs. It posits that individuals' behavioural intention to use a new system is determined by two beliefs: perceived usefulness, defined as the extent to which a person believes that using an IT will enhance his or her job performance and perceived ease of use, defined as the degree to which a person believes that using an IT will be free of effort. It further theorizes that the effect of external variables (e.g., design characteristics) on behavioural intention will be mediated by perceived usefulness and perceived ease of use. TAM has received widespread acclaim and is said to consistently explain over 40% of variance of an individuals' intention to use an IT and actual usage. However, TAM assumes that when someone forms an intention to act, that they will be free to act without limitation. In practice constraints such as limited ability, time, environmental or organisational limits, and unconscious habits will limit the freedom to act so TAM practitioners often compensate for this by introducing additional or alternative belief factors, and/or by examining antecedents and moderators of perceived usefulness and perceived ease of use [11].

2.2 The Unified Theory of Acceptance and Use of Technology

The unified theory of acceptance and use of technology (UTAUT) was first developed by [6] and builds upon elements of eight other models (theory of reasoned action, the technology acceptance model, the motivational model, the theory of planned behaviour, a model combining the technology acceptance model and the theory of planned behaviour, the model of PC utilization, the innovation diffusion theory, and the social cognitive theory). The basic model for assessing information technology acceptance defines the use of a system (a) by the intentions to, and (b) by the individual reactions on, using it. According to its authors, UTAUT provides a useful tool for managers needing to assess the likelihood of success for new technology introductions and helps understand the drivers of acceptance in order to proactively design interventions. In UTAUT, there are four constructs that play a significant role as direct determinants of user acceptance and usage behaviour: Performance Expectancy, Effort Expectancy, Social Influence, Facilitating Conditions; and four moderators: Gender, Age, Voluntariness, Experience.

2.3 The Business Model

The Business Model proposed by [5] outlines 6 core decision variables, namely the value proposition, the target customer segment, the core competency, the competitive positioning, the economic model and the investment model. These decision variables are the inputs to increasingly specific levels of decision-making such as the overarching rules level decisions to foundation and proprietary decisions. By using the Business Model, digital transformation stakeholders can assess, plan and monitor BCT adoption within supply chains and focus resources on areas identified as weak by the model assessment.

2.4 Dynamic Capabilities

Teece et al. [4] define dynamic capabilities as the ability to integrate, build, and reconfigure internal and external competencies to address rapidly changing environments. Building, integrating or configuring Blockchain databases as part of a supply chain transformation can be measured as a dynamic capability. One main argument of this theory is that even if resources do not directly lead the firm to a position of superior sustained competitive advantage, they may nonetheless be critical to the firm's longer-term competitiveness in unstable environments if they help it to develop, add, integrate, and release other key resources over time.

2.5 Resourced-based View of the Firm

The resource-based view (RBV) argues that firms possess resources, a subset of which enable them to achieve competitive advantage, and a subset of those that lead to superior long-term performance. Resources that are valuable and rare can lead to the creation of competitive advantage. That advantage can be sustained over longer time periods to the extent that the firm is able to protect against resource imitation, transfer, or substitution. In general, empirical studies using the theory have strongly supported the resource-based view.

2.6 Diffusion of Innovations Theory

The Diffusion of Innovations Theory (DOI) sees innovations as being communicated through certain channels over time and within a particular social system [8]. Individuals are seen as possessing different degrees of willingness to adopt innovations and thus it is generally observed that the portion of the population adopting an innovation is approximately normally distributed over time. Breaking this normal distribution into segments leads to the segregation of individuals into the following five categories of individual innovativeness (from earliest to latest adopters): innovators, early adopters, early majority, late majority, laggards. Members of each category typically possess certain distinguishing characteristics. The actual rate of adoption is governed by both the rate at which an innovation takes off and the rate of later growth. Low-cost innovations may have a rapid take-off while innovations whose value increases with widespread adoption (network effects) may have faster late-stage growth. Innovation adoption rates can, however, be impacted by other phenomena. For instance, the adaptation of technology to individual needs can change the nature of the innovation over time. In addition, an innovation can impact the adoption rate of an existing innovation and path dependence may lock potentially inferior technologies in place. Research using DOI has consistently found that technical compatibility, technical complexity, and relative advantage (perceived need) are important antecedents to the adoption of innovations.

3. Supply-Chain Transformation in Industry 4.0 and the Circular Economy

3.1 Industry 4.0

According to [12] Industry 4.0 represents the fourth industrial revolution after the mechanization, electrification and computerization of production environments. The manufacturing industry in particular; is set to be completely revolutionised by the increase in automation and digitisation of the industrial shop floor. The concern that streamlined robots all talking to each other via central distributed and coordinated main frames would oust the regular line worker hasn't come to pass. Organisations leading the way in automation and robotics such as Tesla are also major employers and believe that robotics and automation can be powerful tools only when they are properly designed, applied and implemented plus used to supplement rather than replace human skills, knowledge and intuition. This proof of complementary existence highlights the positives of having a blockchain mechanism supporting such production environments. This is important because I4.0 technology may affect the manufacturing industry in a similar way the smartphone technology did for the consumer world by broad behavioural and social changes.

The reason the manufacturing industry is so important is that the digital supply chain models used in modern manufacturing plants are conceptually the same for in-house supply chain as for a broader supply chain that include outside actors. Most organisations adhere to a broader quality framework such as ISO 9001 which will guide their supply chain require requirements and the requirements of those seeking to play a role in that supply chain. The benefits of Digital supply chain (DSC) include cost-effectiveness of services and value-creating activities that are advantageous to many actors in the ecosystem, including firms and their suppliers, employees and customers. Digitization has touched upon all aspects of businesses, including supply chain and operating models. Today, technologies such as RFID, GPS, and sensors have enabled organizations to transform their existing hybrid (combination of paper-based and IT-supported processes) supply chain structures into more flexible, open, agile, and collaborative digital models. Unlike hybrid supply chain models, which have resulted in rigid organizational structures, inaccessible data, and fragmented relationships with partners, digital supply chains enable business process automation, organizational flexibility, and digital management of corporate assets. In order to reap maximum benefits from digital supply chain models, it is important that companies internalize it as an integral part of the overall business model and organizational structure. Localized disconnected initiatives, and silo-based operations pose a serious threat to competitiveness in an increasingly digital world. A holistic approach to digital transformation of supply chain, starting with a digital strategy and a digital operating model will set the direction for integrated execution.

3.2 The Circular Economy

In March 2020 the European Commission published the new circular economy action plan.[1] This plan, presented under the European Green Deal and in line with a new industrial strategy, focuses on proposals for a more sustainable, waste conscious and consumer empowered environment. The European Parliament defines the circular economy to

[1] New_circular_economy_action_plan.pdf (europa.eu).

be a "model of production and consumption, which involves sharing, leasing, reusing, repairing, refurbishing and recycling existing materials and products as long as possible". The circular economy and the push towards sustainability and consumer empowerment in general will increase the demand for Blockchain databases along certain supply chain lines and within certain industrial segments. The circular economy plan specifically mentions Blockchain and argues that by building on the single market and the potential of digital technologies, the circular economy can strengthen the EU's industrial base and foster business creation and entrepreneurship among small to medium enterprises. By using Blockchain technologies the collaboration and sharing of massive amounts of non-personal data called for in the plan can be realised. Indeed, the plan calls for innovative models powered by digital technologies, such as the internet of things, big data, blockchain and artificial intelligence and states that this digital transformation will not only accelerate circularity but also the dematerialisation of the economy and make Europe less dependent on primary materials.

As traditional supply chains are still in the process of evolving from paper-based labelling systems and documentation; the challenge for many organisations is how to transfer to digital supply chains—the awareness of the benefits and the desire to change is often there but the know-how is lacking. In larger organisations the organisational structure is often characterised by functional and geographic silos which do not share information openly.

3.3 Industry 4.0 and CE barriers for Supply Chain Performance

Technology related to I4.0, like the cyber-physical system, IoT, Bigdata, etc., have the capabilities to stay connected and provide critical information throughout the life cycle of products [13]. The convergence of CE and technologies as a smart-circular strategy may reduce the implementation gap of CE through smart remanufacturing, smart reuse, smart recycling, and smart maintenance. [14] reported that the adoption of I4.0 for sustainable supply chain poses organizational, strategic, technological, ethical and legal threats and highlighted 18 critical barriers faced by the sustainable manufacturing sector citing the lack of global standards and data sharing protocols, lack of government support and policies, and financial constraints as the most significant issues restricting the acceptance of I4.0. [15] highlighted the cultural, technological, market and regulatory challenges obstructing I4.0 adoption. [16] pointed out the challenges encountered by industries to achieve I4.0 while fulfilling the objectives of environmental protection. Their study emphasized the importance of technology integration to enable the capabilities of reuse, remanufacturing and recycling. Barriers were highlighted in terms of the cultural aspect, the economic aspect, and the technological and legal aspects. [17] identified the I4.0 barriers against CE adoption through discussion with experts and literature review. Their analysis indicates that "process digitalization," "infrastructure standardization" and "semantic interoperability" are the dominating barriers that can impact the integration of I4.0. Sharma et al. [18] propose that the execution I4.0 supply chains is facing issues due to lack of technology and technique, poor government policies, and lack of user awareness. Kumar et al. [19] highlighted the environmental, government/political, infrastructural, financial, technological, and legal challenges faced by the food supply chain to implement sustainability. Yadav et al. [1] identified supply chain issues such as lack of availability of financial support, technological and human resources, conflicts among sustainability policies, poor management commitment for adoption of sustainability and free trade provisions. They also highlight regulatory uncertainty or the lack of government regulation and public perception or lack of trust among stakeholders as critical barriers against implementing blockchain in supply chains. [20] described the idea of CE as a strategy to improve the ecological aspect of supply chains by enhancing the overall quality of the operation. Also advising to integrate I4.0 and CE to improve the productivity and supply chain security issues. These studies highlight some of the known issues facing a digital transformation project and while BCT has many potential benefits, the implementation barriers remain there as they would for any new and innovative solution.

4. Building the Case for BCT Introduction

Each year organizations are faced with huge holes in their profits due to counterfeit goods. According to a study by the OECD, imports of counterfeit and pirated goods are worth nearly half a trillion dollars a year, or around 2.5% of global imports. Hoping to improve transparency, traceability and immutability in supply chains and to preserve data integrity is of paramount concern with 81 of the world's top 100 public companies by market capitalization using BCT, 27 of them having a fully functioning live product, according to blockchain research firm Blockdata. Another study by MarketsandMarkets estimates that the global blockchain supply chain market will exceed $3.3 billion by 2023 and Gartner forecasts that the business value generated by blockchain reach $176 billion by 2025 and $3.1 trillion by 2030. Despite the obvious advantages BCT brings in terms of and the predicted future growth of the technology [21], show that present blockchain technology-enabled supply chain system (BCTeSCS) efforts are more-oriented toward improving operational-level capabilities (information sharing and coordination capabilities) than strategic-level capabilities (integration and collaboration capabilities). These operational and strategic-level capabilities alongside BCTeSCS deliver several supply chains performance outcomes such as quality compliance and improvement, process improvement, flexibility, reduced cost and reduced process time. However, outcomes may vary by industry type based on their uncertainties and given the nascent state of BCT, accessibility to primary data about ongoing BCTeSCS efforts is limited.

4.1 How the Blockchain Works

The blockchain is a distributed database or a giant, global spreadsheet that runs on millions and millions of distributed computers. Anyone with internet access can view and change the underlying code. It's a true peer to peer network and doesn't require intermediaries to authenticate or to settle transactions, instead it uses state-of-the-art cryptography to record structured information in a global distributed database. Blockchain has since its inception become trusted as an immutable, unhackable distributed database of digital assets.

4.2 What are the Benefits of Blockchain in Supply Chain Management?

Given the desire of organisations to have enhanced transparency and accurate asset tracking capabilities, BCT coupled with the ability to program business logic using smart contracts provides a suitable mechanism to enhance their supply chain efficacy. Current issues with supply chain operations such as mislabeling, illegal sourcing, counterfeiting, etc., are easily addressed by the implementation of public, private, or hybrid blockchains that will bring traceability, transparency, and accountability to the movement of goods and commodities. The technology can be applied to logistics to make business processes more efficient and to cut costs from supply chain infrastructure. In the future and with the Internet of Things capabilities, organisations such as banks won't be able to settle trillions of real-time transactions between things and a blockchain-settlement system will be needed underneath—something which is likely to factor strongly in the future use and development of BTC databases.

The benefits of Blockchain Technology in Supply Chain Management include:

- Products are tracked accurately
- Tampering with products is not possible
- Improved transparency between supply chain stakeholders
- Ability to isolate problems and solve them
- End consumers don't have to deal with a counterfeit assets

- Enhanced produced verification and authentication
- Business process improvements and cost reductions

4.3 Assessing the Supply Chain

Creating a vision for the supply chain provides a company with reference points for the transformation process and is informed by a comprehensive assessment of the supply chain's business and technical requirements and current capabilities. As blockchain gains momentum, companies should keep observing the players in their industry who have begun experimenting with blockchain. Blockchain benefits greatly from network effect; once a critical mass gathers in a supply chain, it is easier for others to jump on board and achieve the benefits. Companies could pay attention to other stakeholders in their supply chain and competitors for indication of timing to develop a blockchain prototype. In assessing supply chain suitability, it must be remembered that supply chains contain complex networks of suppliers, manufacturers, distributors, retailers, auditors, and consumers. A blockchain's shared IT infrastructure would streamline workflows for all parties, no matter the size of the business network. Additionally, a shared infrastructure would provide auditors with greater visibility into participants' activities along the value chain. The main benefit of blockchain technology in supply chains for many organisations is the trust that's embedded in the system. In life-critical industries this can be a game changer. Blockchain has the potential to drive cost-saving efficiencies and to enhance the consumer experience through traceability, transparency, and tradability.

4.4 Horizontal or Vertical Supply Chain Actors

Each company's transformation approach will need to reflect its circumstances, difficulties, opportunities, and goals. Decisions must be made on access rights to private blockchains, e.g., most businesses use both horizontal and vertical integration, choosing the one that is most appropriate for them at set times. Whether to include and/or incorporate suppliers and distributers and to what level are all internal decisions that need to be made by the project champion.

4.5 How Supply-Chain Capabilities and Technologies have Evolved

The low rate of supply-chain digitization has much to do with the capabilities of the technologies that companies have had available until recently. Supply-chain management was one of the first business functions to undergo substantial technology upgrades, as developers created applications to take advantage of data generated by ERP systems. Those applications largely focused on improvements in three areas: streamlining transactional activities such as those involved in end-to-end planning, supporting major operations such as warehouse management, and sharpening the analysis on which decisions are based. What these technologies didn't yet provide, though, were transformative capabilities for supply-chain management: linking and combining cross-functional data (for example, inventory, shipments, and schedules) from internal and external sources; uncovering the origins of performance problems by delving into ERP, warehouse-management, advance-planning, and other systems all at once; or forecasting demand and performance with advanced analytics, so planning can become more precise and problems can be anticipated and prevented.

What's distinctive about the newest digital technologies is that they can integrate better methods for collaboration into a company's processes and prevent a company from regressing to its previous, less effective methods. As companies prepare to transform their supply chains with digital technologies, they need to envision the business and technical capabilities they want and plan to develop those capabilities in tandem.

4.6 Blockchain in Industrial Supply Chains

Blockchain technology is a logical solution to many industries' supply chain issues. The healthcare industry is starting to move to Blockchain databases where transparency and traceability are important. Tracking active pharmaceutical ingredients during the manufacturing process is difficult and faces increased challenges from the widespread and lucrative counterfeit drug operations around the globe. Blockchain's immutability provides a basis for traceability of drugs from manufacture to end consumer, identifying where the supply chain breaks down. There is potential not only to reduce the $200 billion in losses each year but also to increase public safety and prevent some of the estimated one million deaths per year from counterfeit medicine. At government levels throughout the world, mandates are being put in place to improve traceability of government vendors. The United States Congress enacted the Drug Quality and Security Act (DQSA) on November 27, 2013, which mandates unit-level track and traceability for pharmaceuticals by 2023. Dozens of other industries are aiming to meet similar goals to improve the accuracy and overall efficacy of their supply chains. Glaxo Smith Kline (GSK) and several other firms to achieve the regulatory standards by deploying blockchain-based supply chain systems. [22] show that the adoption of CE in supply chains using IoT would improve food security by increasing visibility. The study by [23] demonstrated the potential benefits of the use of technology in supply chains and show that a BTC-enabled supply chain would bring significant changes in traditional practices of food manufacturing and food retailing. So, the implementation use cases are growing and the evidence is beginning to highlight the generic adoption issues faced across all industries.

5. How Theories can Guide Adoption

Although BCT will play a vital role in the digital transformation strategy for enterprises and industry experts speak of value chains, enabled by blockchains, having extended digital platforms for their entire omni-experience ecosystems. What is not so clearly understood, is how this transformation will take shape. BCT is still a nascent technology, thus its limitations and improvements are continuously being discovered and developed. What are the barriers to adoption and how can change management strategists help with the transition? We will see how some of the theories previously discussed can help. By categorising the adoption process into 7 categories we show how the relevant theory can help with this aspect of the process.

5.1 Budgeting

Budgeting is the cornerstone of any change management project and implementing a BCT as part of a supply chain solution is no different. The

Table 2. The BICSVIC Framework.

Challenge	Description	Prescription	Guiding Theory
Budgeting	Unknowns and agile needs	Rapid execution and flexible mechanisms	The Business Model Dynamic Capabilities
Improvement	Continuous Innovation Customer Focus	Reinforcement of transformation themes	Diffusion of Innovations RBV
Control	Data Protection Privacy Security	Collaboration, trust and testing before piloting	The Business Model Dynamic Capabilities RBV
Support	Technical Ability	Skilled talent acquisition Clear digital plan	RBV Dynamic Capabilities
Vision	Corporate Buy In	Strategic alignment	UTAUT
Integration	Assessment of needs	Planning to align with strategy	TAM, RBV
Culture	Workforce Change Management	Sustainable change and collaboration	TAM; UTUAT Diffusion of Innovations

resources must be there to do the implementation correctly. Budgeting for any software development project can be difficult and understanding the requirements and technical challenges involved in a Blockchain implementation makes it difficult to budget accurately upfront. As with any dynamic environment, an agile approach to budgeting is needed. The recommendations for a Blockchain implementation are to set performance goals that support the vision of the organisation. Setting performance goals requires a company to gauge its current performance and then determine achievable improvements. Goals can be defined in terms of capital, and cost measurements. A company that aims to reduce lost sales by a specific amount, for example, would need corresponding supply-chain performance goals. These goals can help bound the project budget and will guide the expenditure on the project.

5.2 Improvement

Digital technology can make customer experiences better by giving supply-chain managers more control and providing customers with unprecedented transparency: for example, track-and-trace systems that send detailed updates about orders throughout the lead time would be very welcome in many instances. Through innovative and continuous improvements, a digital supply chain can help a company strengthen its business model (for example, by expanding into new market segments) and collaborate more effectively with both customers and suppliers (for example, by basing decisions on information that is automatically pulled from upstream customers' or downstream suppliers' ERP systems).

5.3 Control

While solutions exist, including private or permissioned blockchain and strong encryption, there are still cybersecurity breach concerns that need to be addressed before the general public will entrust sensitive data to a blockchain solution. Furthermore, businesses can maintain more control over outsourced contract manufacturing. Blockchain provides all parties within a respective supply chain with access to the same information, potentially reducing communication or transfer data errors. Less time can be spent validating data and more can be spent on delivering goods and services—either improving quality, reducing cost, or both.

5.4 Support

Radiofrequency identification (RFID), 2D barcode, and near field communication (NFC) are used today to link to physical product. However, to ensure the flow of information, all steps of the supply chain and all products will have to be tagged digitally, requiring an overhaul in supply chain practices. Digitally enabled supply chains have talent requirements that can be quite different from those of conventional supply chains. At least some supply-chain managers will need to be able to translate their business needs into relevant digital applications. One key question for managers is can we attract, develop, and retain the "digital native" talent needed to run and transform the supply chain

5.5 Vision

The biggest problems, though, have to do with governance. Any controversy that you read about today is going to revolve around these governance issues. This new community is in its infancy. Unlike the Internet, which has a sophisticated governance ecosystem, the whole world of blockchain and digital currencies is the Wild West. An effective transformation depends on a creative, forward-looking concept for the future supply chain. This means thinking about the outlook for the company, amid the pressures and trends that influence its competitive situation, as well as the changing expectations of its customers. Ultimately, the supply-chain vision should be aligned with the company's strategic goals. While the need for such alignment has always existed, what's new is that both the strategic goals and the vision now must account for the pressures and opportunities that companies face in an increasingly digitized economy.

5.6 Integration

Blockchain solutions require significant changes to—or complete replacement of—existing systems. Organisations need to develop a long-term plan to identify transition requirements for systems required to support blockchain adoption. The integration processes need to be clearly defined and well understood by everyone who is involved in them. This may be needed to ensure the architecture, operating model, and business strategy operate and scale across current established open standards (e.g., ISO 20022, BIAN7) and connect to next-generation third parties such as digital currency marketplaces.

5.7 Culture

Blockchain represents a significant shift to a decentralized network, which requires the buy in of its users and operators. Does the organisation's culture and organizational model encourage experimentation, innovation, and continual improvement? If not, then change managers can use theories to help socialize the idea of implementing blockchain and collaborate with stakeholders prior to implementation to minimize excessive cost or adoption risks.

6. Conclusion

BCT has demonstrated its value and is an exciting and evolving growth area. There are showstoppers such as the energy that's consumed to run blockchain databases. Another issue is that this technology is going to be the platform for a lot of smart agents that are going to displace a lot of humans from jobs and those jobs need to be replaced. That being said, BCT has a huge potential that is currently being realised in the marketplace. Organisations not currently operating with BCT are juggling with the knowledge of its potential versus the costs of implementation. As the technology evolves, the implementation costs and payback periods will start to make more economic sense and it is expected that more and more organisations will begin to use it. While the decision to implement a digital transformation strategy involving a BCT integration is a management level decision, the end users and all vested stakeholders will benefit from the knowledge gained from previous implementation programmes. Using empirically validated theoretical frameworks to navigate through adoption barriers will prove useful for some on this journey.

References

[1] Yadav, G. 2020. A framework to overcome sustainable supply chain challenges through solution measures of industry 4.0 and circular economy: An automotive case, *Journal of Cleaner Production*.

[2] Lezoche, M. 2020. Agri-food 4.0: a survey of the supply chains and technologies for the future agriculture. *Computers in Industry, Elsevier*, 2020, 117: 103187.

[3] Davis, F. 1989. Perceived Usefulness, Perceived Ease of Use, and User Acceptance of Information Technology. *MIS Quarterly*. 13: 319–340.

[4] Teece, D. J., Pisano, G. and Shuen, A. 1997. Dynamic capabilities and strategic management. *Strat. Mgmt. J.*, 18: 509–533.

[5] Clohessy, T. 2019. (De)Mystifying the Information and Communication Technology Business Model Concept. *International Journal of Networking and Virtual Organisations*.

[6] Venkatesh, V., Morris, M. G., Davis, G. B. et al. 2003. User Acceptance of Information Technology: Toward a Unified View. *MIS Quarterly*, 27: 425–478.

[7] Wernerfelt, B. 1984. A resource-based view of the firm. *Strategic Management Journal*, 5(2), 171–180.

[8] Rogers, E.M. 1995. Diffusion of Innovations. 4th Edition, the Free Press, New York.

[9] Ma, Q. and Liu, L. 2004. The Technology Acceptance Model: A Meta-Analysis of Empirical Findings. *Journal of Organizational and End User Computing*, 16: 59–72.

[10] Fishbein, M. and I. Ajzen. 1975. Belief, Attitude, Intention, and Behavior: An Introduction to Theory and Research. Reading, MA: Addison-Wesley.

[11] Wixom, B. H. and P. A. Todd. 2005. A Theoretical Integration of User Satisfaction and Technology Acceptance. *Information Systems Research*, 16: 85–102.

[12] Kagermann, H. 2013. Umsetzungsempfehlungen für das Zukunftsprojekt Industrie 4.0—Abschlussbericht des Arbeitskreises Industrie 4.0.

[13. Alcayaga, A., M. Wiener and E. G. Hansen. 2019. Towards a framework of smart-circular systems: An integrative literature review. *Journal of Cleaner Production*, 221: 622–634.

[14] Luthra, S. and Mangla, S. K. 2018. Evaluating challenges to Industry 4.0 initiatives for supply chain sustainability in emerging economies. *Process Safety and Environmental Protection.*

[15] Kirchherr, J., L. Piscicelli, R. Bour, E. Kostense-Smit, J. Muller, A. Huibrechtse-Truijens and M. Hekkert. 2018. Barriers to the Circular Economy: Evidence From the European Union (EU), *Ecological Economics*, 150: 264–272.

[16] Liboni, L. B., Liboni, L. H. and Cezarino, L. O. 2018. Electric utility 4.0: trends and challenges towards process safety and environmental protection. *Process Safety and Environmental Protection*, 117: 593–605.

[17] Rajput, S. 2019. Industry 4.0 – challenges to implement circular economy Benchmark *Int. J.*

[18] Sharma, Y. K., Mangla, S. K., Patil, P. P. and Liu, S. 2019. When challenges impede the process for circular economy-driven sustainability practices in food supply chain Manag. *Decis.*, 4: 995–1017.

[19] Mangla, S. K., Sharma, Y. K., Patil, P. P., Yadav, G. and Xu, J. 2019. Logistics and distribution challenges to managing operations for corporate sustainability : study on leading Indian diary organizations. *Journal of Cleaner Production*, 238: 117620.

[20] Tseng, M. L., Chiu, A. S., Chien, C. F. and Tan, R. R.. 2019. Pathways and barriers to circularity in food systems Resources. *Conservation Resources*, 143: 236–237.

[21] Nandi, M. 2020. Blockchain technology-enabled supply chain systems and supply chain performance: a resource-based view. *Supply Chain Management*, 25.

[22] Gupta, S., H. Chen, B. T. Hazen, S. Kaur and E. D.R. Santibañez Gonzalez. 2019. Circular economy and big data analytics: A stakeholder perspective, Technological Forecasting and Social Change, 144: 466–474.

[23] Klerkx, L. and D. Rose. 2019. Dealing with the Game-Changing Technologies of Agriculture 4.0: How Do We Manage Diversity and Responsibility in Food System Transition Pathways? *Global Food Security*, 24.

9

Assimilation of the Blockchain

Exploring the Impact of Blockchain Technology on Supply Chain Management

Trevor Clohessy

1. Introduction

Blockchain is a method of storing data in such a manner that it is difficult or impossible to alter, hack, or defraud it. A blockchain is a digital log of transactions that is copied and distributed throughout the blockchain's complete network of computer systems. Blockchain technology is also referred to as distributed ledger technology (DLT). The blockchain was initially published in 2008 when people were first introduced to Bitcoin, a prominent cryptocurrency established by an unnamed inventor, or group of inventors, known as Satoshi Nakamoto; as a result, Nakamoto is also acknowledged as the blockchain's creator. The voyage of blockchain, on the other hand, can be traced back to 1982 when Berkeley programmer David Chum created Blind Signature technology, an untraceable payment method that divorced a person's identity from their transaction. Currently, blockchain is being used as a foundational technology across a range of industries. The primary use cases currently for blockchain technologies include cryptocurrency payments, tracking items along supply chains, non-fungible tokens (NFTs) and copyright management. Supply chains represent a natural setting for blockchain deployments and digital supply chain transformation. For instance, from a financial services perspective, the maturation of blockchain technologies, which

Atlantic Technological University, Renmore, Galway, Ireland.
Email: trevor.clohessy@atu.ie

has enabled the disintermediation of traditional centralized financial service actors/custodians (e.g., banks) has resulted in the emergence of new decentralized financial (DeFi) business models. By utilizing smart contracts on a blockchain, DeFi provides financial instruments without the use of middlemen such as brokerages, exchanges, or banks. People can use DeFi platforms to lend or borrow money, bet on asset price fluctuations using derivatives, trade cryptocurrencies, insure against risks, and earn interest in savings accounts. Financial services that were formerly sluggish and vulnerable to human mistakes are now automated with DeFi smart contracts and safer, thanks to code that anybody can read and evaluate. For example, Binance is a digital asset marketplace that uses the world's leading cryptocurrency exchange's matching engine and wallet technologies to deliver a quick, safe, and trustworthy platform to buy and sell cryptocurrencies internationally.

While blockchain adoption levels across industries are increasing, there are still challenges which are preventing organizations from leveraging the potential of blockchain technologies in supply chains. Specific challenges include a lack of blockchain legislation, and regulation, technology maturity, a lack of blockchain skills and so on. While existing literature has examined the benefits and challenges of blockchain adoption in supply chain contexts using a range of adoption frameworks and lenses (TOE, TAM, Innovation theory) no study has examined them using an innovation theory assimilation perspective (Gallivan, 2001) which posits that technology adoption is a two-stage process that begins with 'primary adoption' (e.g., an organizational choice to embrace an innovation) and ends with 'secondary adoption,' which includes individual adoption and six organizational assimilation phases. Using innovation assimilation theory as a lens for investigating the (i) acceptance, (ii) routinization and (iii) infusion of blockchain technology. The objectives of this study are to examine the following research question:

How does blockchain technology assimilate into supply chain processes from acceptance, routinization, and infusion perspectives?

The next section provides the background for the following study.

2. Background

2.1 Conceptualizing Blockchain Technology

Blockchain technology can be defined as a decentralized digital ledger. Each digital ledger records transactions between entities in blocks. The data in each block is secure, distributed, and anonymized [1]. Blockchain technology was first used to secure the Bitcoin cryptocurrency [2] and came to global renown in 2015 when the technology was featured in The Economist magazine [3] and once again at the end of 2017 when Bitcoin reached a peak value of $20,000. Blockchain is expected to achieve mainstream adoption in multiple industries within the next decade [4]. Furthermore, the blockchain market is expected to grow by 65% in the next five years from $3 billion in 2020 to $39.5 billion in 2025 [5]. Core to many definitions of blockchain are the specific attributes which differentiate the technology from other technologies. Table 1 provides and overview of these six main attributes. First, the concept of access privileges represents a fundamental characteristic of blockchain technology. Akin to private and public cloud computing deployments [6], blockchain can be deployed in a permissioned (private) and permissionless (public) state. In respect to the former, a central authority controls the blockchain network,

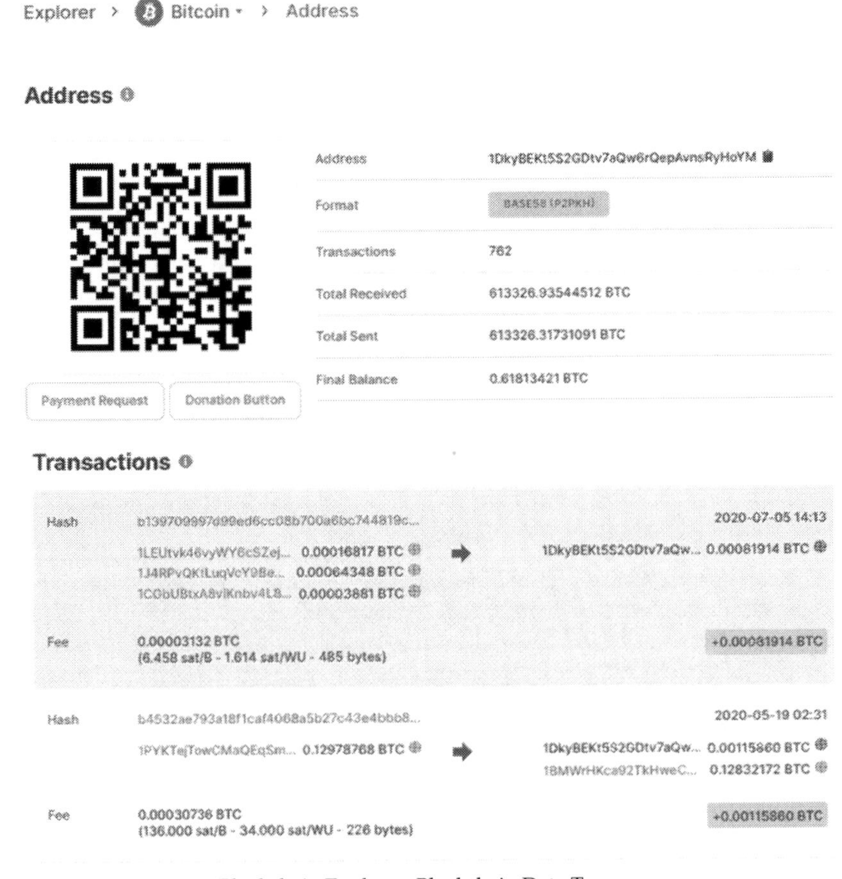

Figure 1. Blockchain Explorer: Blockchain Data Transparency.

Table 1. Blockchain attributes.

Attribute	Definition	Positive	Negative
Access Privileges: Permissioned & Permissionless	Both instances describe the level of public access to data. In public permissioned blockchains, there are no restrictions on reading data. Private permissionless blockchains restrict access to pre-defined users.	Public: Accessibility, and decentralized cooperation. Private: Transaction performance, defined governance structures, innovation speed, data privacy, security, and anonymity.	Public: Transaction performance, governance issues, data privacy, security, and anonymity. Private: Cost, censorship, regulation, and trust.
Immutability	Transactions cannot be altered/deleted once added to the blockchain.	Traceability and business value.	Inflexibility pertaining to the deletion/altering of data.
Transparency	Blockchain facilitates read-only-access to transactions and the inspection of smart contracts contents.	Efficient and accurate record keeping.	Data privacy, smart contract coding errors, compliance with data privacy laws (e.g., EU GDPR).
Programming	Programmable blockchains such as Bitcoin and Ethereum use scripting languages to write digital smart contracts.	The deterministic execution of smart contracts.	Non-programmable blockchains and the complexity of coding real-world contracts into blockchain smart contracts.
Decentralized Consensus	The elimination of a central authority/broker with innovative consensus protocols (e.g., Proof of Stake).	Disintermediation and the creation of new power structures	Proof of work (PoW) energy consumption issues, governance issues and security vulnerabilities.
Distributed Trust	Blockchain does not necessitate high confidence levels in single authorities.	Trust-free systems.	Disintermediation of trusted established stakeholder custodial processes can be difficult to replicate.

while the latter represents an open network where any member can view and write data onto the blockchain. This capability is restricted in a permissioned scenario. Examples of permissionless blockchains include Bitcoin, Ethereum, Cardano. Many enterprise deployments of blockchain technologies use permissioned access privileges for security and privacy reasons [1]. The next characteristic refers to the immutability of the blockchain. That is that once data is entered into a block it is indelible and unalterable. There are obvious benefits and negatives inherent to this characteristic. Emerging blockchain innovations are coming onto the market such as off-chain blockchains which allow transactional data to be edited and/or deleted [7].

Data transparency is the next characteristic of blockchain technology. The ability of all parties in a blockchain network to read the data on the blockchain represents a fundamental new standard for trust [3]. All network participants can access the transactions and holdings of public blockchain addresses by using a blockchain explorer. For example, Figure 1 provides an example of the transactional history for a specific blockchain address which is provided by the blockchain explorer. As can be seen, the explorer provides an overview of the complete transactional history of this specific blockchain user while also ensuring anonymity.

2.2 *The Impact of Blockchain Technology on Supply Chain Management*

Prior to discussing the impact of blockchain technology on supply chain management, it is important to first define the term supply chain management. According to Mentzer et al., [8] supply chain management (SCM) is the "systematic, strategic coordination of the traditional business functions and the tactics across these business functions within a supply chain, for the purpose of improving the long-term performance of the individual companies and the supply chain as a whole". SCM aims to create an efficient and coordinated supply chain via the development of internal and external linkages [9]. Achieving this objective is largely dependent on how effectively organizations use their technological resources and technological capabilities [10,11]. While extant research has focused on how traditional technologies such as cloud computing, RFID, IoT, tracking technologies and so on have impacted SCM business practices, new technological developments are emerging such as blockchain, autonomous robotics and transport, artificial intelligence, drones, digital twins, 3D printing, etc., that are enabling new and innovative SCM business practices [12,13]. Due to the increased attention towards cryptocurrencies in recent years, and in particular Bitcoin, blockchain, the underlying technology that facilitates secure cryptocurrency transactions, has received increasing attention from policymakers, business communities and scholars [11,14].

In fact, many organizations, such as Walmart, Maersk, and Coca Cola, have implemented blockchain in their supply chains. Over the past five years, research has explored how blockchain can benefit SCM business practices (Table 2). As can be seen four blockchain supply chain benefits emerged as being significant: supply chain visibility (N = 14), operational efficiency (N = 11), supply chain traceability (N = 10), supply chain resilience (N = 10), and data security (N = 10). However, many of the benefits have been derived from theoretical/conceptual studies and there is a lack of empirical evidence from actual supply chain settings. Interestingly, considering the recent covid-19 pandemic, blockchain has also been touted as a potential technology solution to ensure supply chain resilience. For example, van Hoek [13] argues that the ability to accelerate information sharing along a supply chain during a global pandemic can enhance its resilience.

Blockchain can minimize global pandemic supply chain risks by "accelerating information sharing and improving visibility into inventory positions and logistics flows… this is particularly the case if the information exchange can shift from partial and sequential (one tier at the time) to more fully and instant" [15]. This ability to disseminate information along a supply chain instantly and rapidly differs dramatically from the traditional sequential and partial distribution of information along a supply chain [11]. It has even been argued that blockchain technologies have great potential to transform disaster relief supply chains through the development and the implementation of blockchain technology solutions in a humanitarian supply chain context [16,17]. The main benefits that could be derived from a humanitarian supply chain setting include swift trust, enhanced collaboration, and increased resilience [14].

2.3 Innovation Assimilation Stages

Assimilation can be defined as "an organizational process that "(i) is set in motion when individual organization members first hear of an innovation's development, (ii) can lead to the acquisition of the innovation, and (iii) sometimes comes to fruition in the innovation's full acceptance, utilization, and institutionalization" [18, p.897]. The assimilation of a technological innovation can often be challenging and is rarely binary [19]. Assimilation theory posits that assimilation may intensify or deteriorate over the course of a technological innovation's adoption journey [20]. Each assimilation stage describes the degree to which the technological innovation permeates the adopting company. Often the causes of innovation success or failure can be minute. While many frameworks have been proposed to understand how innovation assimilation impacts entities [21], one of the most widely cited and used is proposed by Gallivan [19]. Table 3 provides an overview of this six-stage model. The framework comprises two preadoption 'early stages' (e.g., initiation and adoption) and 4 post adoption 'later stages'

Table 2. Supply chain management blockchain benefits.

Authors	Supply Chain Integration	Supply Chain Visibility	Supply Chain Traceability	Supply Chain Innovation	Supply Chain Resilience	Data Security	Data Accessibility	Data Privacy	Operational Efficiency	Enhanced Decision Making
[13]		✓			✓		✓		✓	✓
[14]	✓	✓	✓	✓	✓				✓	
[31]		✓	✓	✓	✓		✓		✓	✓
[32]	✓				✓					
[33]	✓	✓			✓					✓
[34]	✓				✓	✓			✓	✓
[35]	✓	✓	✓		✓		✓		✓	✓
[36]	✓	✓		✓		✓	✓	✓		✓
[37]		✓	✓			✓	✓	✓	✓	✓
[38]	✓	✓	✓			✓				
[39]			✓	✓		✓			✓	
[40]	✓				✓					
[41]		✓	✓	✓		✓		✓		
[42]	✓	✓		✓			✓		✓	✓
[43]		✓	✓		✓	✓	✓	✓	✓	
[44]		✓		✓		✓	✓	✓	✓	✓
[45]		✓	✓		✓	✓		✓		
[46]		✓	✓	✓		✓	✓		✓	
Total:	9	14	10	9	10	10	9	6	11	9

Table 3. Innovation assimilation framework [19].

Stage	Description
Initiation	A match is found for a technological innovation and its application in an organization.
Adoption	A decision is reached to invest resources to adopt the technological innovation.
Adaptation	The technological innovation is developed, installed, and maintained. Organizational members are trained to use the new technological innovation.
Acceptance	Organizational members are induced to use the technological innovation which is now being used within the company.
Routinization	The use of the technological innovation is encouraged as a normal activity and organizational structures are altered to accommodate the technological innovation. It is no longer seen as something out of the ordinary.
Infusion	The technological innovation is used in an elaborate and sophisticated manner. Infusion is categorized in several ways: • **Extended use:** using more features of the technological innovation. • **Integrative use:** using the technological innovation to create new workflow linkages among tasks. • **Emergent use:** using the technological innovation to perform tasks not previously considered possible.

(e.g., adaptation, acceptance, routinization, and infusion). The infusion stage comprises several different facets of technology innovation infusion.

When investigating assimilation stages, a salient consideration is the degree to which we may expect an adopting organization's progression through the assimilation stages to be linear. However, extant research has demonstrated that progression may be non-linear. For instance, [20] argues that a sequential model is more likely to emerge for 'off-the-shelf' technologies in comparison to 'bespoke' technological innovations. Further, [20] suggests that an unfreeze (e.g., initiation, adoption, stages) and unfreeze (e.g., acceptance, routinization, and infusion stages) sequential pattern is likely to emerge. This view, however, has received criticism for failing to consider the degree to which feedback is incorporated into the stages [22]. This six-stage assimilation framework has been used in previous studies that examine technological adoption in supply chains [e.g., 23,24,25]. We also believe it would also be an appropriate lens with which to examine the benefits and challenges associated with the adoption of blockchain technology in a supply chain performance management system context. Although much recent research on technology assimilation has been conducted, the lines of enquiry have focused on the early stages of adoption. As evidence of this unbalanced attention, an analysis was conducted of all technology assimilation supply chain related research published in the Scopus database. The network visualization is displayed in Figure 2.

The prominent keywords emerging from the analysis tend to be 'adoption', 'implementation', and 'diffusion'. Often, these aspects are examined individually and from different theoretical lenses. Furthermore, there is a concentration of research which focuses on the impact of technologies relating to e-commerce (e.g., B2C, B2B), electronic data interchange, and enterprise systems (e.g., ERP, cloud, RFID, tracking technologies). Our study focuses on several of the later assimilation stages (e.g., acceptance, routinization, and infusion). Our review of the literature has identified a research gap which necessitates a need to examine these later assimilation stages with regards to exploring the impact of a new nascent technological innovation called blockchain and its impact on supply chain performance management. Furthermore, these post-adoption stages are deemed to be significant for improving understanding of the various downstream phases of technological innovations [19].

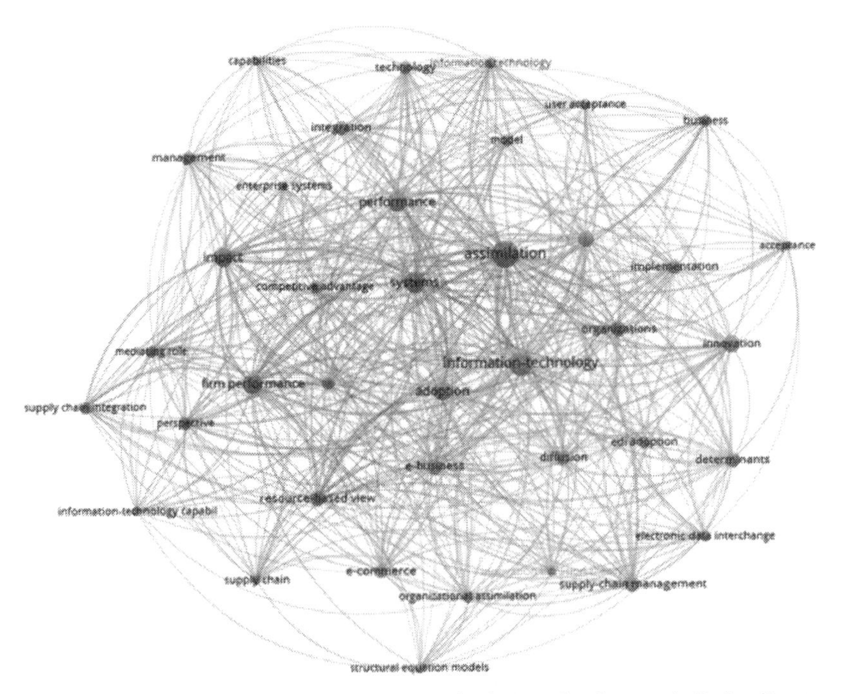

Figure 2. Keyword analysis of the existing supply chain technology assimilation literature.

3. Research Methodology

The objectives of this study are to examine the benefits and challenges of blockchain technology adoption using innovation assimilation theory which is used to determine the (i) acceptance, (ii) routinization and (iii) infusion of blockchain technology. Exploratory case study research was deemed an appropriate research approach for the following reasons. First, there is a scarcity of empirical work around the 'later stages' of blockchain adoption. The case study method "explores a real-life, contemporary bounded system (a case) or multiple bounded systems (cases) over time, through detailed, in-depth data collection involving multiple sources of information... and reports a case description and case themes" [26 p.97]. Second, while previous research has used a combined lens of technological, organizational, and environmental (TOE) theory and force field theory [27], this study is one of the first studies to use the later stage of innovation assimilation theory to investigate blockchain systems in use. A single site case study approach was taken to investigate the assimilation of blockchain technology in supply chain management activities. This approach also facilitates the wider investigation of research questions and aligns the findings with the local conditions of the case [28]. The unit of analysis was the organization and specifically supply chain business processes which had been impacted by blockchain technology. By focusing on these business processes as a unit of analysis it allowed for the emergence of a richer more contextual qualitative narratives and evidence that can be used to decipher how blockchain was accepted, routinised and infused within the organization. Table 4 provides a summary of the

Table 4. Case Study Background.

Company Summary	**Case A**
Industry	DeFi
Company Size *	Medium
Blockchain System Primary Function	Cloud service provider specializing in blockchain node, staking and protocol provision services.
Team size	300+
Location	Distributed
Company Blockchain Development Experience	5 years

case study organization which was used to explore this study's research question. As can be seen Case A are predominantly involved in the provision of blockchain infrastructure for Decentralized Financial (DeFi) Services.

4. Results

In the context of Case A's blockchain supply chain, Table 5 summarizes the later innovation assimilation stage framework elements of acceptance, routinization, and infusion. We have extended the infusion stage to include two further dimensions based on the analysis of the data: diffused use

Table 5. The assimilation stages of blockchain supply chain adoption.

Assimilation stages	Supply chain adoption assimilation
Acceptance	A commitment to using blockchain technology was made in 2016 as a black-box technology or in some bespoke manner
Routinization	A proof-of-concept project in 2016 paved the way for a full roll out of Case A's node infrastructural outsourcing project. Following the roll out of blockchain training workshops and educational initiatives this strategy was embraced by all of Case A's supply chain stakeholders.
Infusion	Blockchain technology was not only used routinely but was also used in a comprehensive and sophisticated manner, which is highlighted by the following aspects: • **Extended use:** Case A's initial use case was providing node infrastructure for customers. Their strategy extended to DeFi products and services such as staking and yield management services. • **Integrative use:** Case A created a nuanced data analytics platform which not only enabled customers to get real-time data insights from their blockchain supply chain processes but also enabled Case A to analyze their supply chain stakeholder's data to provide better services. • **Emergent use:** As blockchain technology matured the company started providing DeFi services which were not in their original strategic scope. • **Diffused use:** As Case A's supply chain stakeholders' numbers started to increase so too did its customer requirements from educational and tailored data insights perspectives. Consequently, the company started to implement additional features to their data analytics platform. They also started to cater for individual enterprise customer bespoke blockchain technology requirements. • **Entrenched use:** Case A's use of blockchain node infrastructural technology is embedded deep into the core functionalities of the organization.

and entrenched use. From an *acceptance* perspective, Case A committed to using blockchain technology in 2016. This stage of assimilation involved a company accepting a technology 'by the book' or 'tailoring it' specifically to their needs. It was noted that blockchain technology back then was very much a black-box technology with limited configurability. There were also only a limited number of DeFI cases on the market. Most frustratingly, blockchain technology was a relatively nascent and unknown technology which resulted in challenges pertaining to sourcing employees with the requisite skills and competencies needed. However, Case A were unswayed by these constraints and remained committed to modifying the technology to suit their DeFi needs. Next, to identify if blockchain technology had reached the *routinization* stage, there must be evidence that the technology is being used regularly or frequently as part of the software development process. Our analysis of Case A deemed that these criteria had been met. As can be seen from Table 5, following the trialing of a proof-of-concept project, blockchain technology was deployed throughout the company. Expert development teams were created to deal with specific blockchain protocols (e.g., Bitcoin, Ethereum, Dogecoin) on which nodes could be created for companies. To overcome the challenges pertaining to sourcing employees with backgrounds in blockchain technologies, Case A created a bespoke internal education program which trained new recruits in the fundamentals of blockchain technologies (e.g., nodes, protocols, staking, DeFi, etc.). This education program was then rolled out to Case A's customers and supply chain partners which further emphasized their embracing of the technology whereby it was not considered something novel or out of the ordinary. Once it had been identified that blockchain technology was being routinely used and embraced by Case A, the study moved onto the next *infusion* stage to examine if Case A were using blockchain technology in a sophisticated and comprehensive manner. The several phases of the infusion stage which were analyzed included: *extended use, integrative use,* and *emergent use.* However, this study has identified that in the case of blockchain technology supply chain adoption, *infusion* is not limited to these three components. There is also *diffused use* and *entrenched use.* By *diffused use* we mean that as Case A'a supply chain and customers base expanded so to do the external channels through which the technology was used. For example, as Case A's customers sizes increased so too did the customer's requirements for additional features for Case A's blockchain products and services. Generally, *diffused use* can only be achieved once the adopting organization has achieved a significant amount of experience with the technology internally prior to rolling out bespoke features to external partners. By *entrenched use* we mean that Case A have successfully used their initial blockchain node infrastructure to develop patented blockchain products and services which would be challenging to imitate by rival organizations. Additionally, this stage is characterized by the fact that the technology is deeply rooted in all the core functionalities of the organization and are essential to the day to day running of supply chain operations. Finally, it was interesting to note that all the blockchain supply chain management benefits that were identified from the literature in Table 2 were also identified in our analysis of Case A's operations. For example, from a supply chain resilience perspective Case A's advanced blockchain analytics platform can predict shifts in cryptocurrency market dynamics in real time, via DeFi Oracles, to protect their customer staking investments.

5. Conclusion

The popularity of blockchain technologies has been increasing over the last 5 years across industries. However, the manner in which blockchain technologies are used and mature within an organization is under researched. From a theoretical perspective, this study provides four contributions. First, this study is one of the first undertaken to explore blockchain supply chain adoption through an innovation assimilation perspective. Second, it is also one of the first to conceptualize the assimilation processes, by focusing on the latter stages of assimilation, (e.g., acceptance, routinization, and infusion). Our analysis of Case A's use of blockchain technology has not only highlighted how blockchain assimilates within their supply chain processes but it also enabled us to extend the assimilation concept to include two new infusion stages from the perspective of blockchain adoption. Next, this extended assimilation lens which has been developed in this study can be used as constructs for quantitatively examining blockchain technology supply chain assimilation in a more in-depth fashion in either the DeFi industry or across other industries. Finally, our analysis of the supply chain literature has identified for main supply chain benefits associated with adopting blockchain technologies: supply chain visibility, operational efficiency, supply chain traceability, supply chain resilience, and data security. As stated earlier, most of the studies which we analyzed were theoretically based. Consequently, our study has confirmed these benefits using an assimilation framework lens.

Our study also has implications for practitioners. The extended blockchain adoption assimilation stage framework and Case A's use of blockchain technologies can help organizations to comprehend how blockchain technologies are adopted and mature within an organization to realize benefits that were previously not thought possible. There are limitations that must be acknowledged in this study. The findings are based on a single case study. Consequently, there are issues regarding the generalizability of the findings. Second, Case A operated in the DeFi industry and as a result their operations are unique to this specific industry which has implications for external validity. However, in-depth analyses of a single case study do not necessarily have to provide wholesale generalizability but instead can provide rich detail and serve as a basis for highlighting how theoretical concepts and frameworks can be applied which served the objectives of this study [29].

Finally, there are a number of avenues which are ripe for further investigation. First, future research could examine the barriers which manifested from adopting blockchain technologies for organizations during the latter stages of assimilation. This would then allow researchers to uncover how organizations overcame these barriers and answer such questions as "What workarounds did they use to facilitate the blockchain adoption process?" Next, this study only looked at the later stages of assimilation. Further research could examine the earlier stages of assimilation (e.g., initiation, adoption, adaption). Insights pertaining to how organizations arrive at their decisions to choose to adopt blockchain technologies, deploy and maintain blockchain technologies, educate employees how to use blockchain technologies on a day-to-day basis and recruit blockchain competent employees, at a time where there is a global skills shortage of these skilled employees, would be welcomed. Finally, further studies could use the extended assimilation framework developed in this study to examine other contexts or industries in which blockchain technology is adopted. Further, it would be interesting to explore cases where the adoption of blockchain technologies failed. This would provide explanations of what contributed to the project's failure and shine a light on the lessons that can be learned from those projects [30].

References

[1] Clohessy, T. and Acton, T. 2019. Investigating the influence of organizational factors on blockchain adoption. *Industrial Management & Data Systems*. ISSN: 0263-5577

[2] Nakamoto, S. 2009 Bitcoin: A Peer-to-Peer Electronic Cash System. Retrieved from: http://www.Bitcoin.org/bitcoin.pdf

[3] Berkeley,J. 2015. The Promise of the Blockchain. The Trust Machine. Retrieved from: https://www.economist.com/leaders/2015/10/31/the-trust-machine on July 2, 2021

[4] Ito, J., Narula, N. and Ali, R. 2017. The Blockchain Will Do to the Financial System What the Internet Did to Media. *Harvard Business Review*, 8.

[5] Markets and Markets 2020. Blockchain Market by Component (Platform and Services), Provider (Application, Middleware, and Infrastructure), Type (Private, Public, and Hybrid), Organization Size, Application Area (BFSI, Government, IT & Telecom), and Region - Global Forecast to 2025. Retrieved from: https://www.marketsandmarkets.com/Market-Reports/blockchain-technology-market-90100890.html?gclid=CjwKCAjwrvv3BRAJEiwAhwOdMzEfR6xkreRPrLu_sEpr-bzX_SDzSX5h4GcPL2bHvIG-YBL0kpTBXBoCaqsQAvD_BwE on July 4, 2020].

[6] Clohessy, T., Acton, T. and Morgan, L. 2018. Contemporary digital business model decision making: a cloud computing supply-side perspective. *International Journal of Networking and Virtual Organisations*, 19(1): 1–20.
[7] IBM 2018. Why new off-chain storage is required for blockchains. Retrieved from: https://www.ibm.com/downloads/cas/RXOVXAPM on June 5, 2021
[8] Mentzer, J. T., DeWitt, W., Keebler, J. S., Min, S., Nix, N. W., Smith, C. D. and Zacharia, Z. G. 2001. Defining supply chain management. *Journal of Business Logistics*, 22(2): 1–25.
[9] Ketchen, D. J., Hult, G.T.M., 2007. Bridging organization theory and supply chain management: the case of best value supply chains. *Journal of Operations Management*, 25(2): 573–580.
[10] Ketchen, D. J. and Giunipero, L. C., 2004. The intersection of strategic management and supply chain management. *Industrial Marketing Management*, 33(1): 51–56.
[11] van Hoek, R. 2019. Exploring blockchain implementation in the supply chain. *International Journal of Operations & Production Management*, 39(6): 829–859.
[12] Zelbst, P. J., Green, K. W., Sower, V. E. and Bond, P. L. 2020. The impact of RFID, IIoT, and Blockchain technologies on supply chain transparency. *Journal of Manufacturing Technology Management*, 31(30): 441–457.
[13] van Hoek, R. 2020. Research opportunities for a more resilient post-COVID-19 supply chain–closing the gap between research findings and industry practice. International *Journal of Operations & Production Management*, 1–15.
[14] Dubey, R., Gunasekaran, A., Bryde, D. J., Dwivedi, Y. K. and Papadopoulos, T. 2020. Blockchain technology for enhancing swift-trust, collaboration, and resilience within a humanitarian supply chain setting. *International Journal of Production Research*, 1–18.
[15] Remko, V. H. 2020. Research opportunities for a more resilient post-COVID-19 supply chain–closing the gap between research findings and industry practice. *International Journal of Operations & Production Management*, 40(4): 341–355.
[16] Thomason, J., M. Ahmad, P. Bronder, E. Hoyt, S. Pocock, J. Bouteloupe and L. Joseph. 2018. Blockchain – Powering and Empowering the Poor in Developing Countries. In Transforming Climate Finance and Green Investment with Blockchains, edited by Alastair Marke, 137–152. Cambridge, MA: Academic Press.
[17] Ramadurai, K. W. and S. K. Bhatia. 2019. Disruptive Technologies and Innovations in Humanitarian Aid and Disaster Relief: An Integrative Approach. In Reimagining Innovation in Humanitarian Medicine, edited by Krish W. Ramadurai and Sujata K. Bhatia, 75–91. Cham: Springer.
[18] Meyer, A. D. and Goes, J. B. 1988. Organizational assimilation of innovations: A multilevel contextual analysis. *Academy of Management Journal*, 31(4): 897–923.
[19] Gallivan, M.2001. Organisational adoption and assimilation of complex technological innovations: development and application of a new framework. *Database for Advances in Information Systems*, 32: 51–85.
[20] Cooper, R. B. and Zmud, R. W. 1990. Information technology implementation research: a technological diffusion approach. *Management Science*, 26: 123–39.
[21] Saga, V. L. and R. W. Zmud. 1994. The nature and determinants of IT acceptance, routinization, and infusion. Levine L, ed. Diffusion, Transfer and Implementation of Information Technology, 67–86
[22] Orlikowski, W. J. and Hofman, J. D. 1997. An improvisational model for change management: the case of groupware technologies. *Sloan Management Review*, 38: 11–21.
[23] Wu, L. and Chuang, C. H. 2010. Examining the diffusion of electronic supply chain management with external antecedents and firm performance: A multi-stage analysis. *Decision Support Systems*, 50(1): 103–115.
[24] Basole, R. C. and Nowak, M. 2018. Assimilation of tracking technology in the supply chain. *Transportation Research Part E: Logistics and Transportation Review*, 114: 350–370.

[25] Simões, A. C., Soares, A. L. and Barros, A. C. 2020. Factors influencing the intention of managers to adopt collaborative robots (cobots) in manufacturing organizations. *Journal of Engineering and Technology Management*, 57: 1–15.
[26] Creswell, J. W. 2013. Research design. Qualitative, quantitative and mixed method approaches (3rd ed.). London: Sage
[27] Kouhizadeh, M., Saberi, S. and Sarkis, J. 2020. Blockchain technology and the sustainable supply chain: Theoretically exploring adoption barriers. *International Journal of Production Economics*, 107831.
[28] Eisenhardt, K. M. and Graebner, M. E. 2007. Theory building from cases: Opportunities and challenges. *The Academy of Management Journal*, 50(1): 25–32.
[29] Donmoyer, R. (2000). Generalizability and the single-case study. *Case study method: Key issues, key texts*, 45–68.
[30] Dwivedi, Y. K., Wastell, D., Laumer, S., Henriksen, H. Z., Myers, M. D., Bunker, D., Elbanna, A., Ravishankar, M. N. and Srivastava, S. C. 2015. Research on information systems failures and successes: Status update and future directions. *Information Systems Frontiers*, 17(1): 143–157.
[31] Clohessy, T. and Clohessy, S. 2020. What's in the box? Combating counterfeit medications in pharmaceutical supply chains with blockchain vigilant information systems. In *Blockchain and distributed ledger technology use cases* (pp. 51–68). Springer, Cham.
[32] Martinez, V., Zhao, M., Blujdea, C., Han, X., Neely, A. and Albores, P. 2019. Blockchain-driven customer order management. *International Journal of Operations & Production Management*.
[33] Ivanov, D., Dolgui, A., Das, A. and Sokolov, B. 2019. Digital supply chain twins: Managing the ripple effect, resilience, and disruption risks by data-driven optimization, simulation, and visibility. In *Handbook of ripple effects in the supply chain* (pp. 309–332). Springer, Cham.
[34] Min, H., 2019. Blockchain technology for enhancing supply chain resilience. *Business Horizons*, 62(1): 35–45.
[35] van Hoek, R. 2019. Exploring blockchain implementation in the supply chain: Learning from pioneers and RFID research. *International Journal of Operations & Production Management*.
[36] Zelbst, P. J., Green, K. W., Sower, V. E. and Bond, P. L. 2019. The impact of RFID, IIoT, and Blockchain technologies on supply chain transparency. *Journal of Manufacturing Technology Management*.
[37] Cottrill, K., 2018. The Benefits of Blockchain: Fact or wishful thinking. *Supply Chain Management Review*, 22(1): 20–25.
[38] Francisco, K. and Swanson, D. 2018. The supply chain has no clothes: Technology adoption of blockchain for supply chain transparency. *Logistics*, 2(1): 2.
[39] Kshetri, N., 2018. 1 Blockchain's roles in meeting key supply chain management objectives. *International Journal of Information Management*, 39: 80–89.
[40] Wu, K. J., Liao, C. J., Tseng, M. L., Lim, M. K., Hu, J. and Tan, K. 2017. Toward sustainability: using big data to explore the decisive attributes of supply chain risks and uncertainties. *Journal of Cleaner Production*, 142: 663–676.
[41] Bocek, T., Rodrigues, B. B., Strasser, T. and Stiller, B. 2017, May. Blockchains everywhere-a use-case of blockchains in the pharma supply-chain. In *2017 IFIP/IEEE symposium on integrated network and service management (IM)* (pp. 772–777). IEEE.
[42] Korpela, K., Hallikas, J. and Dahlberg, T. 2017, January. Digital supply chain transformation toward blockchain integration. In *proceedings of the 50th Hawaii international conference on system sciences*.
[43] Gupta, S. S., 2017. Blockchain. *IBM Online (http://www. IBM. COM)*.

[44] Toyoda, K., Mathiopoulos, P.T., Sasase, I. and Ohtsuki, T. 2017. A novel blockchain-based product ownership management system (POMS) for anti-counterfeits in the post supply chain. *IEEE access*, 5: 17465–17477.

[45] Abeyratne, S. A. and Monfared, R. P. 2016. Blockchain ready manufacturing supply chain using distributed ledger. *International Journal of Research in Engineering and Technology*, 5(9): 1–10.

[46] Apte, S. and Petrovsky, N. 2016. Will blockchain technology revolutionize excipient supply chain management?. *Journal of Excipients and Food Chemicals*, 7(3): 910.

10

Blockchain-Based Energy Efficient Supply Chain Management

Bavly Hanna, Guandong Xu, Xianzhi Wang* and *Jahangir Hossain*

1. Introduction

The Energy Internet (EI) is "a network that merges power technology, electrical technology, information technology, and intelligent management technology" [1]. The conventional energy market trading (ET) mechanism and technical framework were created with a centralized power supply. However, as the modern "energy system" development progresses, there is a growing tendency to elucidate the power retail market and persuade distributed renewable energy (RE) to compete in the market.

The state of blockchain technology (BT) is currently evolving day after day, and "distributed ledger" (DL) technology and "smart contract technology" are increasingly being implemented [2]. Distributed energy and microgrid (MG), are all examples of distributed trading "rules", "algorithms", and "processes". They are built to give fundamental technological alternatives for ET in the current scenario, and technical outlines for EI trading systems for the market relied upon BT are presented.

With the benefits it offers, BT has a chance to become one of the most essential digitalization tools in the energy business. BT can serve as a control mechanism with the rise of distributed energy systems and the demand for green energy. The Blockchain's (BC) distributed structure makes it perfectly compatible with smart grids (SG) and MGs,

University of Technology Sydney.
* Corresponding author: Bavly.s.Hanna@student.uts.edu.au

also crucial components of the energy sector's future [3]. Many scholars and practitioners are interested in the structure of distributed energy systems [4].

The supply chain (SC) and BC technologies are natural matches [5]. To begin with, the BC's structure is a type of time-series data which may hold information, similar to the way items circulate in a SC. Second, the information's relatively low frequency changes in the SC overcomes the processing speed limitations of existing BT.

Each transaction on the BC, including "transaction parties", "transaction time", and "transaction content", will be recorded on a block and kept on each node's DL, ensuring the information's integrity, dependability, and transparency. According to Abeyratne and Monfared, DL technology offers the possibility to minimize delay in transactions, operational risk, turbulence in the process, liquidity needs, and others across a range of financial market use cases [6].

Energy management (EM) has emerged as a critical tool for achieving both economic and environmental goals in a win-win situation. EM minimizes the company's energy expenditures, resulting in a drop in production. It also contributes to environmental well-being by developing innovative, low-polluting technology. Firm and SC sustainability can thus be addressed by deploying a well-designed EM system [7].

Concerns about the impact on the global environment and problems with energy security have heightened the importance of energy efficiency (EE) in recent years. Since energy is used in a variety of industrial processes and at various levels of the SC, implementing EE measures is a difficult undertaking. EE adoption necessitates a complete EM programme at both the individual company and SC levels.

The goal of EM is to manage and optimize the usage of energy in various corporate activities. It may be characterized as the wise use of energy resources to maximize earnings, reduce expenses, and maintain a long-term growth trajectory [8]. Traditionally, EM has been researched at the corporate level. When EM principles are scaled up to the SC level, they must be implemented on the shop floor, in the company, and at many other phases of the SC, such as transportation.

The rest of the book chapter is designed as follows: Section 2 is a literature review about SC-based EM and BC-based EM. Section 3 reports the characteristics, types, classification, and implementation of BC in supply chain management. This is followed by a discussion of the BC applications, advantages, and challenges of EM in Section 4 and conclusions in Section 5.

2. Literature Review

BC is attracting the attention of a wide range of sectors. This increased interest is that BT allows apps that previously relied on a trusted middleman to run without it. Nowadays, they could operate without a verification method and yet achieve the same degree of usefulness and dependability. This was not possible previous to the BC's formation. Trustless networks arose as a result of the use of BT. This is possible since BC networks let users to share data without trusting other users [9].

Due to the absence of intermediaries, user transactions rise at a quicker rate. Furthermore, the BC's use of encryption assures that the data is safe [10]. BC is a massive accounting ledger that keeps track of all user transactions. As a result, Internet of Things (IoT) academics and developers are looking for solutions to connect IoT with BC.

For enterprises that deal with the transfer of products between parties, the SC is nowadays a critical area. The problem with this sector is that it is so large that it can cause delays and defaults in products delivery, among other issues. Furthermore, huge distributors require a large number of employees to meet all of the expectations of merchants. Consequently, this could result in severe order processing delays and an increased risk of orders being lost [10].

To meet this problem, businesses have automated all of their processes. As a result, the number of enterprises and distributors in the SC has increased significantly. Nevertheless, as the amount of digital data rises and Internet companies develop, so does the potential of cyber-attacks on their databases. Hackers may want to change, steal, or remove information [11].

2.1 Related Literature on Supply Chain-based Energy Management

Traditional SC transactions may be transformed into safer, legitimate, and digitally controlled and encrypted blocks with less chance of data tampering using BT. Manupati, et al. designed a "consortium BC" structure for a multi-echelon SC network. A consortium BC is a semi-private BC network administered by numerous SC actors who collaborate on a shared platform [12].

Because all transaction data is safeguarded within the consortium framework, the entire environment is shielded from external dangers. Specialized smart contracts may be created to enforce the defined standards and codes of practice, making BCT easier to adopt in green SC management (SCM). These smart contracts automatically evaluate new transactions against specified criteria, and all SC actors follow the smart contractual agreements [13].

For example, smart contracts can include emission or other product quality determinate threshold limitations. If a transaction exceeds the limit, the smart contract will flag the transaction and alert the responsible parties. SC managers can quickly identify pain areas and change systems to address emissions and product quality concerns. As a result of BT, trust is transferred from human agents to computer codes and smart contracts for transaction verification [14].

Researchers have begun to investigate the implications of BCT in the context of green SCM in recent years. Zhao et al. started the study on BCT applications in sustainable water management and addressed the problems and future research possibilities of BCT in sustainable SCs [15].

The possible uses of BC in the energy market are vast. They might have a considerable influence on both procedures and platforms [16]. For example, BC can lower costs and enable new business models and marketplaces, better manage grid complexity, data security, and ownership, and involve prosumers in the energy market as a facilitator for developing energy communities by lowering costs and enabling new business models and marketplaces [17].

BC can improve the energy market system's transparency and trust, ensure responsibility while adhering to privacy regulations, improve direct peer-to-peer (P2P) ET to ensure the smooth operation of the electricity grid. Also, BC provides a framework for more effective utility billing procedures and transactive energy operations, as well as better handling demand response [18].

BT could be applied to issue certificates of origin, such as the production of green energy and RE sources [19], to construct P2P energy transactions schemes [20], and to set up EM schemes for EV [21]. It is also worth noting that BC is seen as a decarbonization facilitator, allowing the energy sector to shift toward more decentralized energy sources [22].

As the popularity of BC-related technologies grows, so does the demand for energy to power the system. In the past five years, global power consumption predictions for BC-based technology have risen from 2500 to 7670 megawatts [23]. Future technological advancements will significantly influence energy demand, resulting in increased carbon dioxide emissions. According to PwC's BC and cryptocurrency specialists, the entire power requirement for BC-based application servers is 22 terawatt-hours (TWh) each year [24].

That is approximately the same as the total power usage of a country like Ireland. Google, for example, consumes around 5.7 TWh of energy to run its global operations [25]. Another concerning issue is that miners increase their energy use fivefold each year, causing a massive rise that shows no signs of abating.

BT is projected to improve "transparency" and "accountability" of SC, that enables for further flexible "value chains" [26]. In three areas, BC-based solutions have the potential to change SCs: "visibility", "optimization", and "demand" [27]. BC could be used in logistics to identify counterfeit goods, decrease the processing of paper loads, and make provenance tracking easier [28]. Without the need of intermediaries, BC allows buyers and sellers to exchange directly [29].

Moreover, the adoption of BC-based apps in SC networks has been shown to safeguard security [30]. In order to counteract information asymmetry, it leads to more strong contract management systems between third and fourth-party logistics (3PL, 4PL) [31]. It enhances tracking procedures and ensures traceability [32]. It improves information management across the SC. [33]. It strengthens IP protection [34]. Finally, it enhances customer service. [35].

2.2 Related Literature on Blockchain-based Energy Management

In 2008, Nakamoto identified BC as an evolving technology with lots of potential [36]. The first use of BC was as the foundational technology for a cryptocurrency, the so-called "Bitcoin". The tamper-resistance and security of the BC is maintained by mathematics and cryptography [37]. In a P2P network, BT may provide a trust link, allowing all nodes (e.g., firms, consumers, etc.) in the network to put their faith in one another and carry the business forwards.

Apart from financial applications, it is commonly employed in SG technology, IoT, SCM, and data management, among other things [38]. Because of these benefits, we may be able to use BT to address the NEC's collaboration and information sharing issues. To put it another way, all of a driver's charge information may be stored on the BC, which is unchangeable and unaffected by tampering. As a result, secure infrastructure for sharing information on charging in real time might be built and embraced by all businesses.

The BC is "a distributed database that is tamper-resistant, decentralized, and transmitted P2P" [39]. It is a new technology that has the potential to revolutionize a number of established sectors [40]. Apart from its use in Bitcoin, BT protects the accuracy of financial accounting data in the banking sector [41]. Many sectors, such as "decentralized energy resources", SCM, and "carbon emission trading", might benefit from the adoption of BC as it develops and technology improves [42].

There are three sorts of BCs: public BC, private BC, and consortium BC, each with its own set of characteristics. Anyone may join the BC as a new point and execute activities in the public chain (for example "transactions" or "contracts"). The private chain, on the other hand, is a permission BC in which only the approved nodes may join. The consortium BC is a blend of public and private BCs in which only pre-selected nodes may execute accounting operations rather than all nodes participating in the public BC [43]. As a result, organizations are increasingly adopting the consortium BC, which is now being used in a variety of real-world applications.

BC has the potential to transform the energy sector by facilitating transparent, secure, and efficient electrical energy transactions. BC also assists in the development of transactional energy systems, which let scattered agents to trade and interact directly in a decentralized and flat trading market. Splitting the "trading process" into two main stages: the "call auction stage" and the "ongoing auction stage" enables a decentralized and distributed accounting mechanism to match the current requirements of participants' dispersed demands in the energy marketplace.

Energy systems might benefit from the deployment of a BC peer-to-peer network, which would save both energy and money. It enables the democratic connection of various energy sources, guaranteeing that customers have access to low-cost, high-quality energy at all hours and in all locations [44]. This significantly influences society since in many parts of life, energy is required. BC can guarantee that energy is adequately consumed, provided, and allocated in an efficient and transparent way.

On the demand side, "integrated demand response" (IDR) has been shown to improve system operational flexibility and energy usage efficiency by optimizing the operations of "flexible loads", "energy conversion", and "storage equipment" [45]. Zhao et al. proposed an energy transaction method relied upon BT to use the integrated distributed IDR resources fully [46]. Mannaro et al. proposed an energy transaction system based on BT in order to make maximum use of the integrated distributed IDR resources [34]. Mannaro et al. created the "Crypto-Trading Project", emphasizing the importance of BT and smart contracts in administering and controlling the energy market's innovation type [47].

With its "decentralized" and "secure" information sharing platform as its principal characteristic, BT might be an intriguing example of unifying information flow across many parties. The use of BC in the energy sector is still in its early stages, with only a few years of development [48]. However, various research studies, application scenarios, and project endeavors have sparked attention in recent years. Some BC-based frameworks for "energy trading" (ET) for MGs [49] and a market for pricing distributed energy [50] have also been proposed.

The use of information transfers between residential smart meters and "distribution system operators" in Ethereum-based ET has also been suggested [51].

Furthermore, wireless sensor networks and the IoT give alternative solutions to privacy and security challenges through effective data aggregation and the optimization of energy generation and consumption utilizing smart systems that can monitor and interact with one another [52].

The extremely varied energy industry requires a unified SG administration, operation, and control platform. Proper communication and connectivity between the various DG units linked to the main grid contribute to improved SG performance [53]. For monitoring the MG's "voltage" and "frequency", communication between DG units is essential.

3. Blockchain-enabled Supply Chain Attributes

3.1 Blockchain-based Characteristics

The four key characteristics of BC [54] are (1) distributed nodes and storage that are decentralized. (2) It has huge promise in many fields, including banking, computer software, and computer applications, thanks to consensus, smart contracts, and asymmetric encryption. (3) Information and postal economies, as well as investment and securities. (4) the citizen-level MG's shared data structure, creation, and distribution, which may profit from the widened network. Table 1 shows the features of BC that make it distinctive and promising for future industrial uses.

Table 1. Blockchain-based characteristics.

Characteristic	Description
Decentralization	Data could be accessed by multiple organizations.
Scalability	The BC network may be enlarged at any time as new nodes join the network.
Anonymity	Because data is sent across nodes, the individual's identity stays hidden.
Transparent	With network consensus, data is recorded and preserved on the network, and it is accessible and traceable during its full life cycle.
Immutable	To provide immutability, BC offers time-stamps and restrictions.
Untrustworthy	The term "untrustworthy" refers to the fact that the nodes in the BC do not need to trust one other, but just the BC system and its trading mechanism. Because each participating node may become a supervisor under the BC consensus method, there is no need to be concerned about fraud.
Smart contracts	It is a simple electronic software that assists with contract execution.

Due to its promising features and performance in trustless contexts, BT is being hailed as a game-changing invention. It's a time-stamped set of entries saved on a decentralized DL system. There is no central command and control authority, and each node maintains its own digital ledger. All BC records are publicly verifiable and accessible to any node in the network. Following successful transaction validation via a consensus method, a block is formed and appended chronologically. Each block is linked to the preceding one by a unique cryptographic hash, making it unchangeable in nature.

By using a consensus process, the public ledger is updated and synced across all nodes. For transaction validation, the consensus process assures that all participating nodes agree. "Proof of Work" (PoW), "Proof of Stake" (PoS), "Proof of Authority" (PoA), and "Delegated Proof of Stake" (DPoS) are some of the consensus techniques available. The consensus technique is chosen based on the needs of the network; nonetheless, PoW is considered to be more secure than alternative consensus algorithms.

In BC, there are two options. The first is permissionless BC, in which nodes may join the network without requiring any permission, also known as public BC. The second method is a permissioned BC, sometimes known as a private BC, in which access control is used to limit the number of nodes that may participate. Only the authorized nodes in a private BC may read and write data.

With the help of an immutable record, cryptocurrency, as well as smart contract deployment, BC has enabled a unique platform enabling safe energy transactions without trustworthy actors in a dispersed network [55]. For transaction validation, block creation, and hash chain building over the blocks, the agents use a consensus protocol [56].

However, in classic BC applications, such as bitcoin, the commonly utilized proof-of-work consensus system consumes a lot of energy and has delayed transaction confirmation [36]. As a result, it cannot be modified on the permissioned energy BC. As a result, resolving the security issues associated with EV charging in South Carolina remains an open and critical topic.

In an SC made up of networked smart houses, Liang et al. suggested an effective service finding technique for protecting privacy [57]. Zhang et al. analyzed common SG incentive schemes and used a theoretical contract approach to investigate the "cloud-based" "vehicle-to-vehicle" (V2V) ET procedure [58]. Tushar et al. investigated an inducement game-based method for dispersed renewable EM in SC to enhance the operator's gain and reduce overall cost in ET [59].

3.2 Blockchain-based Classification

BC networks are classified in various ways in the current literature [13]. These categories are defined based on the degree of network centralization, network administration, and permissions as (i) "public", (ii) "private", and (iii) consortium "federated".

Public BC is open to the public, and anybody may access data, submit transactions, and "mine" it. As a new user or node miner, anybody could join a public BC (permissionless). Moreover, all parties are capable of carrying out activities like trades and contracts. As a result, the public BC is seen as a totally decentralized presence, as no entity has to influence over it. Bitcoin and Ethereum, as we all know, are examples of public BCs [36,60].

In private BC, the write right is controlled by an organization or institution, and nodes participating in it will be severely limited. In private BCs, that, like consortium BCs, are categorized under the permissioned BC. It's common to provide a whitelist of allowed users with certain capabilities and rights over network operations. Private BC networks can avoid costly PoW techniques since the possibility of Sybil attacks is practically non-existent [62].

A consortium BC is a cross between public and private BCs [62]. In consortium BC, numerous organizations or institutions collaborate on administration. It's worth noting that, unlike the public BC, only the organizations that are part of it have access to data. A wider range of disincentive-based consensus techniques might be utilized.

Even though it provides similar scalability and privacy protection to private BC, the main difference is that instead of a single entity, a group of nodes known as leader nodes is selected to verify transaction processes. This enables for a decentralized design in which leader nodes can grant other users authorization. Table 2 summarizes the major attributes of each BC network in terms of competence, ownership, management of shared information security and consensus mechanisms.

3.3 Blockchain-based Implementation in Supply Chain Management

The integration and coordination of SC functions can be improved with BC. Multiple SC stakeholders can deal with one another without the need for a middleman owing to the BC. As a result, it appears to be best used when there is an issue that affects numerous parties, and each of these parties may gain from solving the problem. This shared value stimulates involvement and facilitates cooperative behavior among participants. Competing organizations can also join the same network thanks to BT.

Table 2. Blockchain-based types.

Features	**Public BC**	**Private BC**	**Consortia BC**
Decentralization degree	Decentralized	Centralization	Weak centralized
Ownership	Public	Centralized	Semi-centralized
Mechanism	Each participant has access to the database, which allows them to save a copy of the transaction and make changes to it	A central authority manages the permissions to access and alter the database	Participants in the consortia
Governance	It is extremely difficult to modify a rule that has already been established	The regulations may be readily modified if the controlling institution makes a choice	It would be simple to amend the regulations by reaching an agreement among consortium members
Data Access	Any user could join the network	Authorized users only could join the network	Limited number of nodes are allowed to join the network
Identifiability	Pseudo-Anonymous	Identified users	Identified users
Privacy	All data are exposed	Privacy is only visible to individual	Flexible and expandable
Transaction proof	Algorithms like PoW, PoS, PoET, and others determine the proof of transaction, which cannot be presented in advance.	The central institution creates proof of transaction such as PBFT, PoA, PoI, etc.	Authentication provides proof of transaction, such as PBFT, PoA, PoI, etc. Transaction verification and block production are carried out according to pre-determined regulations.
Consensus	No permission is required	Permission required	Permission required
Proof of work	Costly	Light	Light
Anonymity	Malicious	Trusted	Trusted

The deployment of BC in SCM is hampered by global economic uncertainty and the lack of a clear framework. Not only for small and medium-sized businesses who are hesitant to engage in BC but also for large corporations because there is no clear structure in place to determine if a SC requires BC implementation or not. The properties of BC are studied and compared to SCM.

Approaches for SCM may vary from one SC to another, or even from one industry to another. BT is gaining popularity as a growing new technology to boost an organization's performance by providing secure and real-time information transmission [63]. BC is, without a doubt, the most up-to-date technology that could be utilized to assist SCM techniques by boosting the integration of all SCM operations [64].

However, it is critical to comprehend the many capabilities provided by BC and how these features might affect various SC activities [65]. If BC characteristics have an influence on the majority of SC processes, it is possible that operational performance will improve as a result. End-to-end visibility and traceability, decentralization, improved data security, decision making, knowledge sharing, end-to-end integration, and effective management are the key emphasis areas for BT in the field of SC [66].

The setup of BC-backed SC systems like Origin Chain is being researched [67]. "Origin Chain" has already established a traceability system that is safe, transparent, and adaptable to regulatory requirements that are always changing. It keeps raw sensitive data off-chain and hashes data on-chain by separating raw information and hash data [68]. Large corporations, like Toyota, have suffered significant losses in the past as a result of the absence of traceability of paint locations along the SC process.

Giants like "Accenture", "Alibaba", and "Walmart" are attempting to use BC in SC, B2B e-commerce, and other sectors in order to combat fraud, ease traceability, and make transactions easier, with an emphasis on high-value items [69]. As a result, the aim is to restructure the SC into one that is scalable and dynamic in response to market developments.

SC players value data sharing security because business transactions frequently involve extremely sensitive commercial information. A permissioned BC might be especially useful in the SC since it provides better privacy, auditability, and operational efficiency [70]. Actors have the ability to encrypt commodity descriptions. They could encrypt sensitive information in a BC SC. They also could and apply commercial guidelines to control access by suppliers or customs [71].

Data integrity and security afforded by BC also guard against fraud and cybercrime. Cybercrime causes data breaches, financial crimes, market manipulations, and intellectual property theft, as well as being a threat to public safety and security. "NotPetY", a recent cyberattack on "Moller-Maersk", the world's largest container shipping line, showed the

SC and logistics infrastructure's vulnerability. The cyberattack caused all of the company's business units to shut down, resulting in a $300 million revenue loss [72].

Despite the fact that a number of variables may be contributing to this vulnerability, installing a decentralized system could assist in risk reduction. A key issue of a centralized system is that if one system is hacked or fails technologically, the entire system may grind to a halt. The BC provides an alternate method of data management that is more resistant to data breaches.

4. Blockchain Applications in Energy Management

4.1 Blockchain-based Applications in Energy Management

BC is being tried in the energy industry for a variety of applications, including security and transparency concerns, as well as process efficiency through the deployment of a decentralized authority idea, resulting in a win-win situation for all parties.

The primary BC-based energy applications are listed in Table 3. BC will enable a trustworthy verification procedure for a trade without requiring identification from a third party in ET applications at the wholesale [73] or local level, such as P2P ET [74]. A typical global BC infrastructure can also make cross-border energy transactions more seamless. Prosumers will be able to engage in the local energy market using BT, which has the potential to make transactions faster, simpler, and less expensive than a typical centralized energy system [75].

BC is also being trialed in electric vehicle (EV) charging facilities. It will give EV drivers access to all charging sites by building a network of EVs and charging stations and a simple payment and settlement procedure for all parties involved. Some energy businesses are also looking at using BC in SC and value chains to improve visibility and prevent asset loss from

Table 3. Blockchain-based applications in energy management.

Application	Description
Decentralized Storage and ET Systems	It allows communities to have more control over the energy sources they use.
P2P Energy Trading	It enables local distributed energy providers to sell their electricity to clients who are prepared to pay the appropriate price.
EV	Vehicle owners may trade electricity by sharing their private EV chargers with others who need them using BT.
Carbon Emission Trading	Companies trade pollution permit.

production to consumption. Project funding is also being looked into, with BC being used to increase payment transparency and liquidity.

The energy market is increasingly changing to a distributed market, with decentralized storage and control in the power grid. The solar photovoltaic business, for example, is becoming increasingly popular. However, when a large number of prosumers develop, how to manage them has become a challenge we must address. For example, when the number of players in the energy market grows, it becomes more difficult to keep their huge data.

Many issues will arise as a result of storing vast volumes of data in a central management organization. The data volume will grow as the number of participants grows, and the accompanying storage cost will rise as well. It will be impossible for other parties to obtain past data if all data is held in one organization, and openness would be tough to enforce.

Furthermore, if central organizations are attacked or hacked, all data is in danger of being exposed or deleted. Voltage regulation, on the other hand, is critical in a decentralized power grid since it affects the grid's operating stability. However, it appears that dealing with these issues will be challenging for the typical centralized administration technique.

Wu et al. created a hybrid BC architecture for EI data storage [17]. The utilization of a combination of public and private BCs to construct an intelligent, efficient, and secure EI management mechanism that incorporates the benefits of both distinguishes this system. It is capable of storing data in a secure and efficient manner. Even though the EI has a high number of participants, it can store and manage its data well.

P2P marketplaces may be built to emphasize the well-being of participants by granting them the greatest amount of personal freedom, financial independence, and privacy possible. Alternatively, they may concentrate on the well-being of the entire community, with participants choosing to share access to a shared resource and working towards a common objective such as total welfare maximization or autonomy [76]. The energy sector is now seeing the emergence of P2P markets [77]. The growth of community-based energy collectives exemplifies this [74].

P2P trading evolved as a modern EM model that permits energy transfer from a prosumer to move to another without the intervention of a "central controller" [78], and is predicted to overcome the limitations of previous systems. A prosumer in need of energy can use a P2P trading platform to benefit from peer prosumers in the network who have energy surplus by purchasing the extra energy at a lower price [59]. Prosumers with extra energy, on the other hand, can make more money through P2P ET than under the feed-in-tariff (FiT) arrangement.

Stevanoni et al. investigate an intriguing notion of energy sharing between peers and MGs [79]. The goal is to represent the varied, often contradictory, intentions of the stakeholders in the enduring development of peer-linked industrial MGs. It combines long-term investment decisions with short-term operation activities using two "game-theoretical frameworks". To show the technical and economic outcomes, the created instrument is assessed on a simulated industrial MG.

MG-to-MG ET mechanisms are P2P ET mechanisms that are primarily focused on addressing the imbalance between energy supply and demand inside MGs by maximizing the use of RE resources. Such procedures lower electricity expenditures for MGs that participate in P2P ET [80].

Despite the stochastic load demand, it is feasible to smooth power generation at the MGs by managing the grid's electricity flow [81]. Incorporating security into P2P trading also permits for the spread of dispersed energy resources and demand flexibility, resulting in a more robust system in both conventional and extraordinary situations [82].

To solve the privacy and transaction security challenges for plug-in hybrid EVs, Kang et al. presented a novel P2P ET model with a consortium BC solution [83]. Aitzhan and Svetinovic introduced a "token-based decentralized ET system" based on multi signatures and anonymous encryption technologies to allow peers to conduct transactions secretly and safely [84]. Li et al. used the consortium BT to secure the distributed ET market and developed a new energy BC system for the industrial IoT [85].

Lower pricing, higher EV performance, quicker charging rates, and the availability, speed, convenience of use, and affordability of charging infrastructure will all play a role in the future rise of EVs. According to Su, et al., if the BC is incorporated into the energy grid, the electric car charging infrastructure would improve. India and the EU, for example, are working to boost the development of an electric car ecosystem [86].

Though, many charging providers face complex payment arrangements, payment methods that are not all the same, limited charge heaps, and erroneous charging cost estimation in the case of a real-time charging application EV. These issues can be solved with BT [87]. BT is mostly utilized in charging station operating platforms, which helps to reduce the inconvenient nature of electric car charging, efficiently manage to charge infrastructure, boost security, and encourage the use of shared batteries and shared energy for a common good [88].

BC can help with climate action by improving carbon emissions trading and promoting RE trade [95]. Consumers can purchase, sell, or trade RE using tokens reflecting energy output or transferrable digital assets in the domain of climate finance innovation on a global scale. BT helps to increase climate finance flows by assisting in the development of P2P financial transactions to promote climate action and guaranteeing that funds are given to projects with higher transparency.

4.2 Blockchain-based Advantages in Energy Efficiency

Improving EE allows a business to provide the same service (for example, product manufacturing, energy transfer from one form to another) while consuming less energy. Furthermore, because customers are becoming more environmentally conscious, EE has the ability to improve a single firm's total performance and create appropriate cost savings as well as additional revenues [89].

Companies within the same SC should cooperate to better serve consumer expectations than their competitors in an environment marked by high globalization, quickly growing technology, and increasingly demanding customers. As a result, in today's business climate, aligning strategic and operational choices throughout the many phases of the SC is a must for gaining and preserving competitive advantages.

Companies may decrease environmental and social problems while improving economic results by working closely with consumers and suppliers. As a result, it's critical to recognize the possible impact of SC interactions on enhancing EE. Most businesses, on the other hand, pay just passing attention to whether their partners use EM systems in their operations.

In an ever-changing business environment, collaboration among SC partners will offer value to each partner and the SC as a whole because risks can be shared, costs can be saved, and lead and reaction times can be decreased [90]. As stated by Jaber and Goyal, SC coordination methods are either centralized, in which a single decision-maker manages the whole SC, or decentralized, in which several decision-makers with competing agendas are engaged, resulting in an inefficient system [91].

Despite the fact that many scholars have lately focused on the rising relevance of EE implementation in operations and SCM [92]. The previous literature has mostly looked at energy-efficient investment decisions made inside a single company from an industrial standpoint. During the previous several decades, there has been a lot of research on EE in the industrial sector, and there have been various literature evaluations.

Biel and Glock look at decision support models that include energy considerations in industrial businesses' mid- and short-term production planning [93]. Schulze et al. conduct a thorough evaluation of existing academic journal papers on industrial EM [94]. Tanaka creates a platform for policy analysis aimed at improving industrial EE and conservation [95].

Pons et al. created a map of implemented technologies for reducing energy usage in manufacturing and linked them to manufacturing company performance [96]. However, it is equally critical to recognize the possible impact of SC interactions on such investment choices [97]. Cosimato and Troisi demonstrated how developing green technologies with a focus on logistic activities might help to improve EE, reduce toxic emissions, boost the use of RE sources, and better manage or reuse trash [98].

It has been established that EE has a considerable favorable influence. Customers would save around $500 billion annually if EE enhancements were implemented, according to the International Energy Agency's EE 2018 report [99]. As BT advances, users will be able to exchange their extra energy. Because they would be able to monetize their excess energy, consumers will be even more motivated to save energy and improve the EE of their houses. As the EE business matures, BT has the potential to enhance significantly overall administrative operations, transparency, cost, and stakeholder confidence.

A multi-agent system (MAS) has successfully tackled several challenges, including sound monitoring in various scenarios [100]. The use of a MAS to solve logistical problems is not a new topic; a MAS is presented as a solution to the logistical challenge [101]. Furthermore, the challenge of distributed computing is another successful use of MAS [102]. As a result, several of the literature-based concepts combine the benefits of BC with MAS. The work of Aitzhan and Svetinovic stands out among a variety of systems that merge BC with MAS [84]. Some research promotes security and privacy in decentralized energy networks by combining both technologies. Yuan and Wang suggest a ride-sharing system that uses MAS and BC [103]. There are several additional BC and MAS applications. Daza et al. suggest a novel BC approach for IoT [104].

SCEMs require data that is relevant to the system and technology advancements in order to map energy usage across the chain. The capacity to digitize and display data from food SC has been made possible by the emergence of a new generation of IT at a breakneck pace [105].

Industry 4.0 technologies, in particular "BC", "cloud computing", "IoT", are progressively applied to different SCs. They make it possible to collect and share data in order to increase automation, product development, and material management efficiency [106].

Such technologies could assist in the decision-making process in SC monitoring by allowing for the study and visualization of energy usage and carbon emissions in on-demand services [107]. IoT, for example, allows for the gathering of multi-source data using sensing technology throughout the SC, employing sensors that record and transfer data via communication technologies (such as 5G or the internet).

Sensors might detect the electricity usage of processing equipment, the refrigerating temperature, the position of items throughout the SC, and more to capture energy use along the SC. This type of data can aid in the tracking of a product's embodied energy and carbon emissions along the SC. Specifically, each product's SC may be traced using BT, and detailed information on energy usage per product can be shared via energy labeling on the items. The most energy-efficient SC sequence may also be released and utilized as a blueprint for future SC reconstruction using BT

4.3 Challenges of Blockchain-based Integration

For each SC, the relevance of integration differs. To improve their operational effectiveness, some SCs only depend on integration-based SC methods. Since SCM procedures have an influence on a variety of key performance indicators (KPIs) such as lead time, cost savings, flexibility, and so on, and these KPIs are used to develop operational performance measurements [108]. The literature, on the other hand, shows that SCM methods have an influence on operational performance.

There are several BC applications in sectors such as commodities traceability, copyright, and SC application scenarios in the overall economy. When it comes to integrating BC-based solutions in the real world, however, there are still a number of substantial hurdles. Because the SC is fragmented, multifaceted, and geographically spread, proposing an effective scheme to assure the security and fairness of the "information", "logistics", and "capital" flow of the SC at the same time is extremely difficult. Furthermore, the viability of such a scheme, as well as its execution are still being worked out [109].

In addition to all of these benefits, there are certain drawbacks to using BC to move to a circular economy. Integrating BT with current traditional information management systems is one of these issues. Due to various obstacles, technological transformations are not possible to achieve swiftly. These difficulties may arise from a variety of factors, including a lack of technology infrastructure, bureaucracy, and human resources. The barriers to BC and EM integration should be emphasized, and top management should be appropriately placed.

While DL technologies present a lot of possible societal gains, however, they also pose several ethical and environmental problems. BC, similar to other technologies, is only a means to increase system efficiency. According to Hawken et al., a fundamental rethinking of the structure of the business is required, inspired by natural capitalism discussion, to guarantee that the use of technology is centered on getting the right things done rather than merely doing something in the correct manner [110].

Focusing just on eco-efficiency without integrating BC into a value-driven business model would be bad for the environment by enabling resource-saving in incorrect business models, where resources are harvested quicker, and incorrect goods reach the market at a cheaper cost. Consider a digital platform that connects customers to merchants that reward them with coins for their environmentally friendly conduct, which they can then use to buy non-green things from the vendors. Another example is a digital platform that improves the efficiency of a regionally distributed SC with greater CO_2 emissions than local supply networks by facilitating shipment tracking.

Another issue is that existing BC protocols are computationally demanding. For example, the Bitcoin network has various scalability issues, including user communication failures, data retention, and linear transaction history, as well as miners competing on transaction verification, resulting in considerable energy loss [111].

Aside from the scale issue, the technology's current limitations include BC's complexity, the need for a decentralized network of users, and unchangeable human errors. Due to the immutable nature of BC, it is crucial to double-check the first information put on BC platforms, as material stored on the platforms is irreversible, and there must be systems in place to avoid human mistakes.

Aside from technological limitations, hurdles to BC deployment include a lack of government regulations and a lack of trust among stakeholders, according to the literature. Other success elements for BC in the SC include data security, quality, accessibility, and documentation [112].

5. Conclusion

The benefits of BC applications are progressively becoming apparent in the sectors of new energy trade, financial services, and energy SCM. BT appears to be a viable alternative for the development of RE [14]. Furthermore, while developed nations dominate fundamental energy BC research due to their mastery of more sophisticated BT, poorer countries are increasingly key players in the global energy BC research system. In terms of publications, China has considerably overtaken the developed nations led by the United States, implying that developing countries may soon overtake developed countries to dominate global energy BC research [113].

The world's energy BC use cases are more prevalent in industrialized countries. This is due to the fact that industrialized nations have relatively full energy infrastructures, such as distributed power grids (MGs), which make energy BCs a viable application environment [114]. The majority of home power supply facilities in developing nations led by China are centralized energy supply systems. This is the polar opposite of BT's decentralized properties, and it stymies the technology's continued development and deployment in the energy sector.

As a result, underdeveloped nations will have a significant issue in establishing energy BCs in the future: how to swiftly create a comprehensive distributed power grid. This directly decides whether developing nations can transform their energy BC research advantages into practical energy BC applications and subsequently support RE development to improve the energy structure.

References

[1] Jaradat, M., Jarrah, M., Bousselham, A., Jararweh, Y. and Al-Ayyoub, M. 2015. The internet of energy: smart sensor networks and big data management for smart grid. *Procedia Computer Science*, 56: 592–597.

[2] Zhu, X., He, Q. and Guo, S. 2018. Application of Blockchain technology in supply chain finance. *China Circulation Economy*, 32(3): 111–119.

[3] Caputo, F., Buhnova, B. and Walletzky, L. 2018. Investigating the role of smartness for sustainability: insights from the smart grid domain. *Sustainable Science*, 13: 1299–1309.

[4] Hirsch, A., Parag, Y. and Guerrero, J. 2018. Microgrids: a review of technologies, key drivers, and outstanding issues. *Renew. Sustain. Energy Rev.*, 90: 402–411.

[5] Eyal, I. 2017. Blockchain technology: transforming libertarian cryptocurrency dreams to finance and banking realities. *Computer*, 50(9): 38–49.

[6] Abeyratne, S. A. and Monfared, R. P. (2016). Blockchain ready manufacturing supply chain using distributed ledger. *International Journal of Research in Engineering and Technology*, 5: 1–10.

[7] Schaltegger, S. and Figge, F. 2000. Environmental shareholder value: economic success with corporate environmental management. *Eco-Management and Auditing*, 7(1): 29–42. Doi:10.1002/(SICI)1099-0925(200003)7:1<29::AIDEMA119>3.0.CO;2-1.

[8] Capehart, B. L., Turner, W. C. and Kennedy, W .J. 2006. Guide to Energy Management (5th ed.). Lilburn, GA: The Fairmont Press.

[9] Chamoso, P., Rivas, A., Martin-Limorti, J. J. and Rodriguez, S. 2018. A hash based image matching algorithm for social networks. *In Advances in Intelligent Systems and Computing*, 619: 183–190.

[10] Li, T., Sun, S., Bolic, M. and Corchado, J. M. 2016. Algorithm design for parallel implementation of the SMC-PHD filter. *Signal Processing*, 119: 115–127.

[11] Cardoso, R. C. and Bordini, R. H. 2017. A multi-agent extension of a hierarchical task network planning formalism. *Advances in Distributed Computing and Artificial Intelligence Journal* (ADCAIJ), Salamanca, 6(2).

[12] Manupati, V. K., Schoenherr, T., Ramkumar, M., Wagner, S. M., Pabba, S. K. and Singh, R. I. R. 2020. A blockchain-based approach for a multi-echelon sustainable supply chain. *International Journal of Production Research*, 58(7): 2222–2241.

[13] Christidis, K. and Devetsikiotis, M. 2016. Blockchains and smart contracts for the internet of things. *IEEE Access*, 4: 2292–2303.

[14] Mengelkamp, E., Notheisen, B., Beer, C., Dauer, D. and Weinhardt, C. 2018. A blockchain-based smart grid: towards sustainable local energy markets. *Computer Science-Research and Development*, 33(1-2): 207–214.

[15] Zhao, G., Liu, S., Lopez, C., Lu, H., Elgueta, S., Chena, H. and Boshkoska, B. M. 2019. Blockchain technology in agri-food value chain management: A synthesis of applications, challenges and future research directions. *Computers in Industry*, 109: 83–99.

[16] Bilal, K., Malik, S. U. R., Khalid, O., Hameed, A., Alvarez, E., Wijaysekara, V., Irfan, R., Shrestha, S., Dwivedy, D., Ali, M., Shahid Khan, U., Abbas, A., Jalil, N. and Khan, S. U. 2014. A taxonomy and survey on green data center networks. *Future Generation Comput. Syst.*, 36: 189–208.

[17] Wu, L., Meng, K., Xu, S., Li, S. Q., Ding, M. and Suo, Y. 2017. In: Democratic Centralism: A Hybrid Blockchain Architecture and its Applications in Energy Internet, in: Proceedings – 1st *IEEE International Conference on Energy Internet*, pp. 176-181.

[18] Kyriakarakos, G. and Papadakis, G. 2018. Microgrids for productive uses of energy in the developing world and blockchain: a promising future. *Appl. Sci.*, Switzerland, 8: (4).

[19] Park, L. W., Lee, S. and Chang, H. 2018. A sustainable home energy prosumer-chain methodology with energy tags over the blockchain. *Sustainability*, Switzerland, 10(3). Parliament, E.

[20] Pop, C., Cioara, T., Antal, M., Anghel, I., Salomie, I. and Bertoncini, M. 2018. Blockchain based decentralized management of demand response programs in smart energy grids. *Sensors*, Switzerland 18 (1).

[21] Huang, X., Xu, C., Wang, P. and Liu, H. 2018. LNSC: a security model for electric vehicle and charging pile management based on blockchain ecosystem. *IEEE Access* 6, 13565–13574. https://doi.org/10.1109/ACCESS.2018.2812176.

[22] World Energy Council and PricewaterhouseCoopers. 2018. The Developing Role of Blockchain. White paper. Executive Summary. https://www.worldenergy.org/wpcontent/uploads/2017/11/WP_Blockchain_Exec-Summary_final.pdf.

[23] Clark, E. C. and Greenley, L. H. 2019. Bitcoin, Blockchain, and the Energy Sector, Independently published. Retrieved from https://www.hsdl.org/?view&did=828211.

[24] PricewaterhouseCoopers. 2016. Blockchain-an opportunity for energy producers and consumers? Retrieved from https://www.pwc.com/gx/en/industries/energy-utilities-resources/publications/opportunity-for-energy-producers.html.

[25] Baraniuk, C. 2019. Bitcoin's energy consumption 'equals that of Switzerland'. Retrieved from https://www.bbc.com/news/technology-48853230.

[26] Kshetri, N. 2018. Blockchain's roles in meeting key supply chain management objectives. *Int. J. Inf. Manage.*, 39: 80–89.

[27] IBM Corporation. 2016. Making Blockchain Real for Business. Explained with High Security Business Network Service. https://www.ibm.com/systems/data/flash/it/technicalday/pdf/Making%20blockchain%20real%20for%20business.pdf.

[28] Tan, A. W. K., Zhao, Y. and Halliday, T. 2018. A blockchain model for less container load operations in China. *Int. J. Inf. Syst. Supply Chain Manage.*, 11(2): 39–53.

[29] Subramanian, H. 2017. Decentralized blockchain-based electronic marketplaces. *Commun. ACM* 61(1): 78–84.

[30] Dorri, A., Kanhere, S. S. and Jurdak, R. 2017. Towards an optimized blockchain for IoT. In: Proceedings IEEE/ACM 2nd *International Conference on Internet-of-Things Design and Implementation*, IoTDI, pp. 173–178.

[31] Polim, R., Hu, Q. and Kumara, S. 2017. Blockchain in megacity logistics. In: IIE Annual Conference. Proceedings, *Institute of Industrial and Systems Engineers* (IISE), pp. 1589–1594.

[32] Lu, Q. and Xu, X. 2017. Adaptable blockchain-based systems: a case study for product traceability. *IEEE Softw.* 34(6): 21–27.

[33] O'Leary, K., O'Reilly, P., Feller, J., Gleasure, R., Li, S. and Cristoforo, J. 2017. Exploring the application of blockchain technology to combat the effects of social loafing in cross functional group projects. In: *Proceedings of the 13th International Symposium on Open Collaboration*, OpenSym.

[34] Holland, M., Nigischer, C. and Stjepandic, J. 2017. Copyright protection in additive manufacturing with blockchain approach. *Adv. Transdiscip. Eng.*, 5: 914–921.

[35] Frey, R. M., Wörner, S. and Ilic, A. 2016. Collaborative filtering on the blockchain: a secure recommender system for E-commerce. *In*: AMCIS 2016: Surfing the IT Innovation Wave, *22nd Americas Conference on Information Systems*.

[36] Nakamoto, S. 2008. Bitcoin: A Peer-to-Peer electronic cash system. https://www.bitcoincash.org/bitcoin.pdf.

[37] Li, J. X., Wu, J. G. and Chen, L. 2018. Block-secure: blockchain based scheme for secure P2P cloud storage. *Inf. Sci.*, 465: 219–231.

[38] Casino, F., Dasaklis, T. K. and Patsakis, C. 2019. A systematic literature review of blockchain-based applications: current status, classification and open issues. *Telematics Inf.*, 36: 55–81.

[39] Aggarwal, S., Chaudhary, R. and Aujla, G. S. 2019. Blockchain for smart communities: applications, challenges and opportunities. *J. Netw. Comput. Appl.*, 144: 13–48.
[40] Morkunas, V. J., Paschen, J. and Boon, E. 2019. How blockchain technologies impact your business model. *Bus. Horiz.*, 62: 295–306.
[41] McCallig, J., Robb, A. and Rohde, F. 2019. Establishing the representational faithfulness of financial accounting information using multiparty security, network analysis and a blockchain. *Int. J. Account. Inf. Syst.*, 33: 47–58.
[42] Choi, T. M., Wen, X., Sun, X. T. and Chung, S. H. 2019. The mean-variance approach for global supply chain risk analysis with air logistics in the blockchain technology era. *Transport. Res.* Part E 127: 178–191.
[43] Huang, X. H., Zhang, Y., Li D. D. and Han, L. 2019. An optimal scheduling algorithm for hybrid EV charging scenario using consortium blockchains. *Future Generat. Comput. Syst.*, 91: 555–562.
[44] Park, L. W., Lee, S. and Chang, H. 2018. A sustainable home energy prosumer-chain methodology with energy tags over the blockchain. *Sustainability*, 10(3): 658.
[45] Li, Y., Wang, C., Li, F. and Chen, C. 2021. Optimal scheduling of integrated demand response-enabled integrated energy systems with uncertain renewable generations: a Stackelberg game approach. *Energy Convers. Manag.*, 235: 113996. Doi: 10.1016/j.enconman.2021.113996.
[46] Zhao, S., Wang, B., Li, Y. and Li, Y. 2018. Integrated energy transaction mechanisms based on blockchain technology. *Energies*, 11: 2412. Doi: 10.3390/en11092412.
[47] Mannaro, K., Pinna, A. and Marchesi, M. 2017. Crypto-trading: blockchain oriented energy market, in *AEIT International Annual Conference*, Cagliari: IEEE, 1–5. Doi: 10.23919/AEIT.2017.8240547.
[48] Wang, Q. and Su, M. 2020. Integrating blockchain technology into the energy sector - From theory of blockchain to research and application of energy blockchain. *Comput. Sci. Rev.*, 37: 100275.
[49] Xue, L., Teng, Y., Zhang, Z., Li, J., Wang, K. and Huang, Q. 2017. Blockchain technology for electricity market in microgrid. In Proceedings of the 2nd International Conference on Power and Renewable Energy (ICPRE), Chengdu, China, 20–23 September 2017; IEEE: Piscataway, NJ, USA, pp. 704–708.
[50] Cheng, S., Zeng, B. and Huang, Y. Z. 2017. Research on application models of blockchain technology in the distributed electricity market. *In IOP Conference Series: Earth and Environmental Science*, IOP Publishing: London, UK, 93: 012–065.
[51] Hussain, S. S., Farooq, S. M. and Ustun, T. S. 2019. Implementation of blockchain technology for energy trading with smart meters. In Proceedings of the Innovations in Power and Advanced Computing Technologies (i-PACT), Vellore, India, 22–23 March 2019; IEEE: Piscataway, NJ, USA, 1: 1–5.
[52] Alladi, T., Chamola, V., Rodrigues, J. J. and Kozlov, S. A. 2019. Blockchain in smart grids: A review on different use cases. *Sensors*, 19: 4862.
[53] Galici, M., Mureddu, M., Ghiani, E., Celli, G., Pilo, F., Porcu, P. and Canetto, B. 2021. Energy Blockchain for Public Energy Communities. *Appl. Sci.*, 11: 3457.
[54] Giungato, P., Rana, R., Tarabella, A. and Tricase, C. 2017. Current trends in sustainability of bitcoins and related blockchain technology. *Sustainability*, 9: 2214. Doi: 10.3390/su9122214.
[55] Novo, O. 2018. Blockchain meets IoT: An architecture for scalable access management in IoT. *IEEE Internet Things J.*, 5(2): 1184–1195.
[56] Dorri, A., Steger, M., Kanhere, S. S. and Jurdak, R. 2017. BlockChain: A distributed solution to automotive security and privacy. *IEEE Commun. Mag.*, 55(12): 119–125.
[57] Liang, X., Zhang, K., Lu, R., Lin, X. and Shen, X. 2013. EPS: An efficient and privacy-preserving service searching scheme for smart community. *IEEE Sensors J.*, 13(10): 3702–3710.

[58] Zhang, K., Mao, Y., Leng, S., Maharjan, S., Zhang, Y., Vinel, A. and Jonsson, M. 2016. Incentive-driven energy trading in the smart grid. *IEEE Access*, 4: 1243–1257.

[59] Tushar, W., Chai, B., Yuen, C., Smith, D. B., Wood, K. L., Yang, Z. and Poor, H. V. 2015. Three-party energy management with distributed energy resources in smart grid. *IEEE Trans. Ind. Electron.*, 62(4): 2487–2498.

[60] Ethereum, https://www.ethereum.org.

[61] Swanson, T. 2015. Consensus-as-a-service: a brief report on the emergence of permissioned, distributed ledger systems, https://allquantor.at/blockchainbib/pdf/swanson2015consensus.pdf.

[62] Zheng, Z., Xie, S., Dai, H. N. and Wang, H. 2016. Blockchain challenges and opportunities: a survey. Work Pap.

[63] Behnke, K. and Janssen, M. F. W. H. A. 2020. Boundary conditions for traceability in food supply chains using blockchain technology. *International Journal of Information Management*, 52 https://doi.org/10.1016/j.ijinfomgt.2019.05.025.

[64] Cole, R., Stevenson, M. and Aitken, J. 2019. Blockchain technology: implications for operations and supply chain management. *Supply Chain Management*, 24(4): 469–483. https://doi.org/10.1108/SCM-09-2018-0309.

[65] Viriyasitavat, W. and Hoonsopon, D. 2019. Blockchain characteristics and consensus in modern business processes. *Journal of Industrial Information Integration*, 13: 32–39. https://doi.org/10.1016/j.jii.2018.07.004.

[66] Feng, Q., He, D., Zeadally, S., Khan, M. K. and Kumar, N. 2019. A survey on privacy protection in blockchain system. *J. Network Comput. Appl.*, 126: 45–58.

[67] Wang, Y., Han, J. H. and Beynon-Davies, P. 2019. Understanding blockchain technology for future supply chains: a systematic literature review and research agenda. *Supply Chain Manage. Int. J.*, 24(1): 62–84.

[68] Shamout, M. 2019. Understanding blockchain innovation in supply chain and logistics industry. *Int. J. Recent Technol. Eng.*, 7(6): 616–622.

[69] Kshetri, N. and Loukoianova, E. 2019. Blockchain adoption in supply chain networks in Asia. *IT Prof.*, 21(1): 11–15.

[70] Weber, I., Xu, X., Riveret, R., Governatori, G., Ponomarev, A. and Mendling, J. 2016. Untrusted business process monitoring and execution using blockchain. *International Conference on Business Process Management*, 9850: 329–347.

[71] Kshetri, N. 2017. Will blockchain emerge as a tool to break the poverty chain in the global South? *Third World Quarterly*, 38(8): 1710–1732.

[72] Milne, R. 2017. Moller-Maersk puts cost of cyber-attack at up to $300m. *Financial Times*. Available online: www.ft.com/ content/a44ede7c-825f-11e7-a4ce-15b2513cb3ff (accessed 12 February 2022).

[73] Enerchain. First European Energy Trade Over The Blockchain. 4 November 2016. Available online: https://enerchain.ponton.de/index.php/11-first-european-energy-trade-over-the-blockchain (accessed on 12 Feb. 2022).

[74] Sousa, T., Soares, T., Pinson, P., Moret, F., Baroche, T. and Sorin, E. 2019. Peer-to-peer and community-based markets: A comprehensive review. *Renew. Sustain. Energy Rev.*, 104: 367–378.

[75] Verma, P., O'Regan, B., Hayes, B., Thakur, S. and JBreslin, G. 2018. EnerPort: Irish Blockchain project for peer- to-peer energy trading. *Energy Inform.*, 1: 1–9.

[76] Einav, L., Farronato, C. and Levin, J. 2016. Peer-to-Peer Markets. *Annu. Rev. Econ.*, https://doi.org/10.1146/annurev-economics-080315-015334.

[77] Zhang, C., Wu, J., Long, C. and Cheng, M. 2018. Peer-to-peer energy trading in a microgrid. *Appl. Energy*, 220: 1–12. https://doi.org/10.1109/TSG.2013.2284664.

[78] Tushar, W., Yuen, C., Mohsenian-Rad, H., Saha, T., Poor, H. V. and Wood, K. L. 2018. Transforming energy networks via peer-to-peer energy trading—the potential of game-theoretic approaches. *IEEE Signal Process Mag.*, 35(4): 2–24.

[79] Stevanoni C, Grve, Z. D., Valle, F. and Deblecker, O. 2018. Long-term planning of connected industrial microgrids: a game theoretical approach including daily peer-to-microgrid exchanges. *IEEE Trans Smart Grid*, PP(99): 1.

[80] Liu, T., Tan, X., Sun, B., Wu, Y., Guan, X. and Tsang, D. H. K. 2015. Energy management of cooperative microgrids with p2p energy sharing in distribution networks. *IEEE international conference on smart grid communications* (SmartGridComm), Miami, FL. p. 410-5.

[81] Fathi, M. and Bevrani, H. 2013. Statistical cooperative power dispatching in interconnected microgrids. *IEEE Trans Sust Energy*, 4(3): 586–93.

[82] Moslehi, K. and Kumar, R. 2018. Autonomous resilient grids in an IoT landscape - vision for a nested transactive grid. *IEEE Trans Power Syst*, PP(99): 1.

[83] Kang, J., Yu, R., Huang, X., Maharjan, S., Zhang, Y. and Hossain, E. 2017. Enabling localized peer-to-peer electricity trading among plug-in hybrid electric vehicles using consortium blockchains. *IEEE Trans. Ind. Informat.*, 13(6): 3154–3164.

[84] Aitzhan, N. Z. and Svetinovic, D. 2016. Security and privacy in decentralized energy trading through multi-signatures, blockchain and anonymous messaging streams. *IEEE Trans. Depend. Secure Comput.*, 15(5): 840–852, Sep./Oct. 2016.

[85] Li, Z., Kang, J., Yu, R., Ye, D., Deng, Q. and Zhang, Y. 2018. Consortium blockchain for secure energy trading in industrial Internet of Things. *IEEE Trans. Ind. Informat.*, 14(8): 3690–3700.

[86] Su, Z., Wang, Y., Xu, Q., Fei, M., Tian, Y. and Zhang, N. 2019. A secure charging scheme for electric vehicles with smart communities in energy Blockchain. *IEEE Internet Things J.*, 6: 4601–4613.

[87] IEA, Global EV Outlook 2018: Towards Cross-Modal Electrification, International Energy Agency (IEA), France, 2018.

[88] Liu, C., Chai, K. K., Zhang, X., Lau, E. T. and Chen, Y. 2018. Adaptive blockchain-based electric vehicle participation scheme in smart grid platform, *IEEE Access*, 6: 25657–25665.

[89] Jayaram, J. and Avittathur, B. 2015. Green supply chains: A perspective from an emerging economy. *Int. J. Prod. Econ.*, 164: 234–244.

[90] Jansen, J. H. 2014. Supply Chain Finance Management. *HAN Business Publications*, Nijmegen, The Netherlands, pp. 1–36.

[91] Jaber, M. Y. and Goyal, S. K. 2018. Coordinating a three-level supply chain with multiple suppliers, a vendor and multiple buyers. *Int. J. Prod. Econ.*, 116: 95–103.

[92] Nguyen, J. Q., Donohue, K. and Mehrotra, M. 2015. The buyer's role in improving supply chain energy efficiency. *SSRN*. Available online: https://ssrn.com/abstract=2564287.

[93] Biel, K. and Glock, C. H. 2016. Systematic literature review of decision support models for energy-efficient production planning. *Comput. Ind. Eng.*, 101: 243–259.

[94] Schulze, M., Nehler, H., Ottosson, M. and Thollander, P. 2016. Energy management in industry - A systematic review of previous findings and an integrative conceptual framework. *J. Clean. Prod.*, 112: 3692–3708.

[95] Tanaka, K. 2011. Review of policies and measures for energy efficiency in industry sector. *Energy Policy*, 39: 6532–6550.

[96] Pons, M., Bikfalvi, A., Llach, J. and Palcic, I. 2013. Exploring the impact of energy efficiency technologies on manufacturing firm performance. *J. Clean. Prod.*, 52: 134–144.

[97] Wu, Z., Ellram, L. M. and Schuchard, R. 2014. Understanding the role of government and buyers in supplier energy efficiency initiatives. *J. Supply Chain Manag.*, 50: 84–105.

[98] Cosimato, S. and Troisi, O. 2015. Green supply chain management. *TQM J.*, 27: 256–276.

[99] International Energy Agency. 2018. Energy Efficiency Analysis and Outlooks to 2040; OECD/IEA: Paris, France, 2018.

[100] Casado-Vara, R., Gonzalez-Briones, A., Prieto, J. and Corchado, J. M. 2018. Smart contract for monitoring and control of logistics activities: pharmaceutical utilities case study. *The 13th International Conference on Soft Computing Models in Industrial and Environmental Applications*, Springer, Cham.z, pp. 509–517.

[101] Casado-Vara, R., Prieto, J. and Corchado, J. M. 2018. How Blockchain Could Improve Fraud Detection in Power Distribution Grid. *The 13th International Conference on Soft Computing Models in Industrial and Environmental Applications*, Springer, Cham, pp. 67–76.

[102] Banerjee, S. and Hecker, J. P. 2017. A Multi-agent system approach to load-balancing and resource allocation for distributed computing. *In*: Bourgine, P., Collet, P. and Parrend, P. (eds.). First Complex Systems Digital Campus World E-Conference. *Springer Proceedings in Complexity*. Springer.

[103] Yuan, Y. and Wang, F. Y. 2016. Towards blockchain-based intelligent transportation systems. *19th International Conference on Intelligent Transportation Systems (ITSC)*, IEEE, Rio de Janeiro, pp. 2663–2668.

[104] Daza, V., Di Pietro, R., Klimek, I. and Signorini, M. 2017. CONNECT: CONtextual NamE discovery for blockchain-based services in the IoT. *Communications (ICC) 2017 IEEE International Conference*, 1–6, 2017, ISSN1938-1883.

[105] Han, Y., Zhou, R., Geng, Z., Bai, J., Ma, B. and Fan, J. 2020. A novel data envelopment analysis cross-model integrating interpretative structural model and analytic hierarchy process for energy efficiency evaluation and optimization modeling: Application to ethylene industries. *Journal of Cleaner Production*, 246: 118965. https://doi.org/10.1016/j.jclepro.2019.118965.

[106] Fisher, O. J., Watson, N. J., Porcu, L., Bacon, D., Rigley, M. and Gomes, R. L. 2021. Multiple target data-driven models to enable sustainable process manufacturing: An industrial bioprocess case study. *Journal of Cleaner Production*, 126242. https://doi.org/10.1016/j.jclepro.2021.126242.

[107] Fisher, O., Watson, N., Porcu, L., Bacon, D., Rigley, M. and Gomes, R. L. 2018. Cloud manufacturing as a sustainable process manufacturing route. *Journal of Manufacturing Systems*, 47: 53–68. https://doi.org/10.1016/j.jmsy.2018.03.005.

[108] Roespinoedji, R., Lisdayanti, A., Farida, S. and Isa, A. M. 2019. Supply chain management practices and operational performance: the mediating role of process control and improvement. *International Journal of Supply Chain Management*, 8(2): 159–167.

[109] Badal, F. R., Das, P., Sarker, S. K. and Das, S. K. 2019. A survey on control issues in renewable energy integration and microgrid. *Prot. Control. Mod. Power Syst.*, 4: 8.

[110] Hawken, P., Lovins, A. B. and Lovins, L. H. 2013. *Natural Capitalism: the Next Industrial Revolution*. Routledge.

[111] Barber, S., Boyen, X., Shi, E. and Uzun, E. 2012. *Bitter to better - how to make Bitcoin a better currency. International Conference on Financial Cryptography and Data Security*. Springer, pp. 399–414.

[112] Yadav, S. and Singh, S. P. 2020. Blockchain critical success factors for sustainable supply chain. *Resour. Conserv. Recycl.*, 152: 104505.

[113] Wang, Q., Su, M. and Li, R. 2020. Is China the world's blockchain leader? Evidence, evolution and outlook of China's blockchain research. *J. Cleaner Prod.*, 264: 121742.

[114] Parino, F., Beiró, M. G. and Gauvin, L. 2018. Analysis of the Bitcoin blockchain: socio-economic factors behind the adoption. *EPJ Data Sci.*, 7: 38.

Index

A

B

C

D

E

F

G

H

Printed in Great Britain
by Amazon

7528799d-f7ac-49c4-a1b9-2f1f75f14e76R01